LEDs for Lighting Applications

LEDs for Lighting Applications

Edited by
Patrick Mottier

ISTE

WILEY

First published in France in 2008 by Hermes Science/Lavoisier entitled: *Les diodes électroluminescentes pour l'éclairage* © LAVOISIER, 2008
First published in Great Britain and the United States in 2009 by ISTE Ltd and John Wiley & Sons, Inc.

ISTE Ltd
27-37 St George's Road
London SW19 4EU
UK

www.iste.co.uk

John Wiley & Sons, Inc.
111 River Street
Hoboken, NJ 07030
USA

www.wiley.com

Library of Congress Cataloging-in-Publication Data

Diodes électroluminescentes pour l'éclairage. English
LEDs for lighting applications / edited by Patrick Mottier.
p. cm.
Includes bibliographical references and index.
ISBN 978-1-84821-145-2
1. Light emitting diodes. 2. Electric lighting--Equipment and supplies. I. Mottier, Patrick.
TK7871.89.L53D5613 2009
621.3815'22--dc22

2009013274

British Library Cataloguing-in-Publication Data
A CIP record for this book is available from the British Library
ISBN: 978-1-84821-145-2

FSC
Mixed Sources
Product group from well-managed forests and other controlled sources
Cert no. SGS-COC-2953
www.fsc.org
© 1996 Forest Stewardship Council

Table of Contents

Foreword

How could we possibly underestimate the achievement of Thomas Edison when he manufactured the first incandescent lamp in October 1879? Humanity had finally succeeded in producing light without combustion, without smell, without smoke. From then on, such light was to be available anywhere and everywhere that a source of electricity existed. 130 years later, this technology is still widespread, though the use of arc lamps has been increasing since the 1950s.

However, a more recently developed technology for the production of light could well consign incandescent lighting to oblivion. Though quite expensive at the moment (but prices are going down every day), it seems to offer a much higher level of performance in all respects: lifespan, compactness, output, color adjustment and ease of dimming. The technology in question is the Light-Emitting Diode (LED).

As a pure product of optoelectronics, the LED came along at just the right time. It was in 2008 that governments round the world made a decision to speed up reductions in energy use and the associated CO_2 emissions. They also decided, more generally, to cut down on the use of resources.

Lighting accounts for 19% of electrical consumption worldwide, which means that it presents a considerable potential for reductions over the next decade. Moreover, LEDs offer a way forward: they can already compete with incandescent lights, and in due course they will rival arc lights.

This opportunity, which is historical in its importance, is closely related to environmental issues. It goes hand in hand with a growing demand for lighting systems that can better satisfy consumer needs and can also entail lifestyle spinoffs. LEDs seem especially adaptable to maintenance devices using sensors, wireless control and power supplied by photovoltaic panels and batteries.

LED technology should benefit from the impressive developments currently taking place in optoelectronics generally, particularly in the fields of mobile phones, computers and TV screens, automobile equipment and advertising panels.

Still, the question remains: "Do we really need such flexible, versatile light sources?"

The almost unlimited possibilities of optoelectronic technologies, with respect to flux dimming and color control, will undoubtedly have repercussions for photobiology and psychology, as regards exposure to light (or other forms of radiation). Technical specifications should no longer involve light that imparts visibility alone, but also light that promotes comprehension, predictability, safety, beauty and emotion. In short, what is now required is light for *life*.

It would be a mistake to underestimate the possible benefits of a light source such as the LED. The prospects of increased luminous efficacy at 120-200 lm/W (i.e. the ratio between light flux and electrical power) make it a front-runner with regard to the production of white light (incandescent, halogen, arc lamps). This would provide a further incentive to develop it in such a way as to take optimal account of environmental considerations.

The present book constitutes a major contribution to the issues involved, given the outstanding quality of the information it provides, and the date of its publication. 2009 is, in effect, the year when LED-based products are finally starting to enter the market on a large scale. A new lighting revolution is underway.

Patrick Mottier has brought together the leading French experts to produce this work, which should provide LED manufacturers and lighting specialists with essential strategic information. It should also prove indispensable to all of those who, for whatever reason, wish to broaden their knowledge of the field.

I'm sure you'll take pleasure in reading it.

Dr Marc Fontoynont

Head of the Building Sciences Laboratory
National Engineering School of State Public Works, Lyon, France
University of Lyon
Vice-President, International Lighting Commission, Vienna, Austria

Introduction

Is the revolution running in lighting? Indeed, the lighting domain is facing a unique technology breakthrough with the emergence of new kinds of devices and luminaries based on semiconductor chips, so-called Light-Emitting Diodes or LEDs. In the beginning, the low power and fluxes of LEDs impeded them from being used for lighting and kept them confined to indicator applications. However, the invention of the blue LED in the middle of the 1980s gave access to white light combining red, green and blue LEDs or by means of partial photoconversion of blue light to yellow light using phosphors. White LEDs in their turn have made ?electronic lighting? accessible, opening new business opportunities to companies. Consequently, labs and companies have focused their efforts on increasing LED efficiencies and fluxes. All these efforts have already led to significant results.

LED-based lighting is a recent phenomenon, but LED deployment in this field of application now seems inevitable. The LEDs have become a credible alternative to incandescent lamps, soon to be banned because of their energy inefficiency. The multiple benefits of LEDs and the continuous increase in their performance allied to the decrease of their manufacturing costs are likely to make them competitive when compared to fluorescent lamps and tubes. Will 2009 mark the true take-off of LED lighting? We do not know yet, but in any case High Brightness white LEDs have become a reality and it is the objective of this book to give to its readers a wide scientific and technological overview of what they are, including all LED manufacturing steps, aspects related to their photoelectric characterization up to characteristics of the light they produce. It also opens up opportunities in organic LEDs.

Chapter 1 starts with a quick history of LEDs, then positions them within the lighting problematic, before setting out issues and challenges still facing High Brightness LEDs. We can consider this chapter to constitute the actual introduction to this book.

The next four chapters describe the successive LED manufacturing steps.

First, Chapter 2 presents the gallium nitride (GaN) material, the semiconductor on which any blue LED is achieved and the problems bound to its epitaxy on so-called "hetero-substrates". Basic techniques of metallo-organic vapor phase epitaxy are presented and finally some results on bulk GaN crystal growth are given.

Chapter 3 is devoted to the junction itself. After a short history of the discovery of blue LEDs, the authors focus on the achievement of the p-n junction in the GaN semiconductor and then on quantum wells inserted into the p-n junction as a way to increase the LED efficiency. Next, they discuss diode optical properties and radiation efficiency before commenting on possible future developments.

Chapter 4 focuses on processing of diode heterostructure wafers to achieve a high efficiency LED. Different LED structures are described and the evolutions of LED design are presented with their advantages and drawbacks. Then, the successive technological processing steps are discussed.

Chapter 5 deals with LED chip packaging. It presents a quick historical review from the first LED devices to today's products. Then, problems specifically bound to High Brightness LEDs and related to thermal management are discussed. It then addresses primary optics issues with light extraction and related materials. Finally, the different characteristics given in the LED technical data sheet are discussed.

Chapter 6 is devoted to LED characterization. It begins by focusing on its photometric aspects, i.e. on their luminous behavior, and then looks at their electrical and thermal characterization.

Lighting by LEDs, however, requires the LEDs to reach a certain level of light quality: this is the subject of Chapter 7. It outlines the fundamentals of white light, before addressing the various ways to produce white light from

LEDs. Finally, the author presents and discusses some recent works on the estimation of the quality of light coming from LED-based sources from a lighting point of view.

Finally, if the very concept of LED is the result of the invention of blue LED on gallium nitride, which now holds pride of place, the emergence of OLEDs, or Organic Light-Emitting Diodes, must be taken into account now. Chapter 8 discusses the technology originally developed for display, considering here its application to lighting. We hope that, by presenting this comprehensive overview, this book will meet the expectations of engineers, teachers and students concerned with the technology that aids breakthroughs and opens up possible sources of innovation in an area, lighting, which until a few years ago we imagined would have technical relative stability.

I would like to express my extreme thanks to the authors for their great contributions to this book.

Patrick MOTTIER

Chapter 1

Light-Emitting Diodes: Principles and Challenges

1.1. History of a revolution in the world of the light sources

"During an investigation of the unsymmetrical passage of current through a contact of carborundum and other substances a curious phenomenon was noted. On applying a potential of 10 Volts between two points on a crystal of carborundum, the crystal gave out a yellowish light [...]".

Henry Joseph Round (Marconi Co, UK) reported a century ago, in 1907, the first known report of the effect of the light emission from a semiconductor material [ROU 07]. Figure 1.1 shows a reconstitution of Captain Round's experiment. The first LED patent was filled in 1929 by a Russian radio technician Oleg Vladimirovich Losev [LOS 29].

Without clear explanation, this discovery was quickly forgotten. It was not until 1962 that Nick Holonyak[1] and S.F. Bevacqua, consultants at General Electric, signed "the official birth certificate" of the first red light-emitting diode [HOL 62] (Figure 1.2).

Chapter written by Georges ZISSIS.

1. In 2002 N. Holonyak received the *National Medal of Technology* from the President of the USA for his invention. In 1995 he also received the "Japan prize" awarded to honor the achievements of people who have contributed to the progress of science and technology and the advancement of world peace and prosperity.

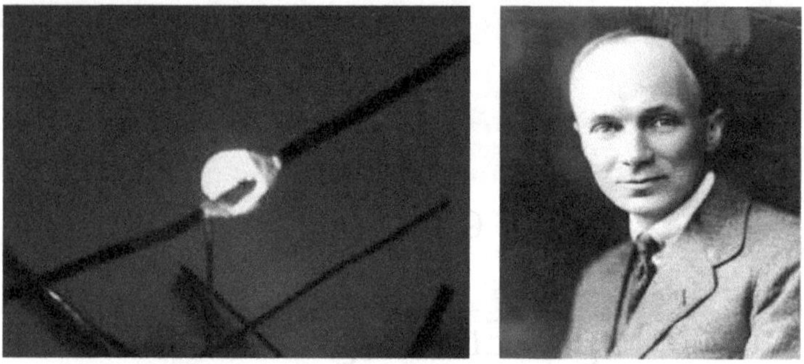

Figure 1.1. *Silicon carbide crystal emitting yellow light under electrical excitation; this picture reproduces the original 1907 experiment of captain H.J. Round (right picture), accidental illustration of light emission from semiconductors*

Figure 1.2. *Professor N. Holonyak (left), inventor of visible light LEDs, shows his demonstrator to his mentor, Professor J. Bardeen, inventor of the transistor (picture taken in 1972, credits IPPO)*

Since then, things have rapidly gained pace. In 1968, the first commercialized LED (Light-Emitting Diode) produced barely 0.001 lm of red light. Nowadays High Brightness[2] white LEDs, generating more than 100 lm, are commercially available[3]. Recently, a record value of 1,100 lm was achieved with devices combining several LED junctions on a chip[4]. This is a true revolution compared to the general trend: in the last 30 years, the

2. High Brightness corresponds to a luminous flux between 50 and 250 lumens. Ultra High Brightness to more than 250 lumens.

3. Luxeon has presented an experimental white LED generating 500 lm.

4. Osram Ostar with six junctions (2006).

luminous flux generated by a single device doubles each 18 to 24 months (a 20-fold increase per decade). In the meantime the price per device also rapidly decreases (a tenfold decrease per decade). This trend, shown in Figure 1.3, was first observed by Roland Haitz and hence is known as "Haitz's Law" [HAI 02].

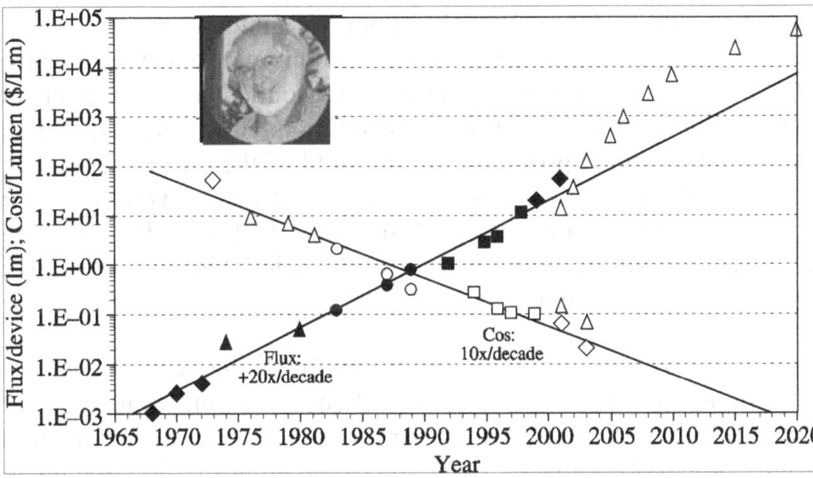

Figure 1.3. *"Haitz's Law" predicts that the luminous flux generated by LEDs increases 20 times per decade whereas the cost decreases 10 times within the same time period (inserted picture: R. Haitz)*

An obvious question can legitimately be asked following these revolutionary results: can LEDs replace the traditional illumination devices (light bulbs and electrical discharge systems)?

The question is worth asking but the answer is not as obvious. Of course, many technological and scientific aspects have to be taken into account, but the most important issue is to evaluate this innovative light source in the light of today's economic and energy saving considerations.

1.2. LEDs and lighting

Electrical lighting has drastically changed everyone's everyday life to a point that one could not live without it. It is estimated that 30 billion light bulbs are in use on the planet, consuming each year a total of

2,650 TWh, that is about 19% of the total electricity produced worldwide[5]. Today, two main technologies dominate the market, incandescence and electrical discharge.

Although these technologies are now mature, the luminous efficiency of the light sources together with their quality of light have not quite reached their limits: there is still room for innovation. The design, optimization and mass production of new and more efficient light sources for lighting are scientific, technical, economic and environmental challenges, but seem to be a sustainable solution. However, despite many scientific and technical progresses in the field of the electrical discharge light sources, the maximum efficiency of these systems has been growing since the 1970s to reach about 100-110 lm/W[6]. LEDs on the other hand, with a steady growth of their luminous efficiencies (Figure 1.4), establish themselves as breakthrough solutions.

If white LEDs could reach 200 lm/W by 2025, they could replace the existing fluorescent lights[7]. The estimated energy saving would be of the order of 1 billion barrels of oil per year (which corresponds to an electricity production capacity reduction of the order of 250 large nuclear plants[8]).

This energy saving would also imply a significant reduction in CO_2 emission, more than 270 millions tons per year. On top of this, the use of white LEDs in lighting systems will also significantly transform this domain because of its numerous advantages compared to traditional lights:

- higher energetic efficiency of the illumination systems;
- longer lifetime and therefore less maintenance;
- size reduction of the equipments;
- higher flexibility and control of the level of light and color variation;

5. As of today, electricity represents approximately 16% of the total energy produced worldwide (20% for industrialized countries); lighting therefore consumes 3% of the energetic resources per year.

6. Only good quality white light sources are considered here.

7. The luminous efficiency is the only criterion considered here, but other technological aspects also have to taken into account.

8. Each "plant" represents 2 TWh/year.

– low power consumption (battery driven autonomous systems, consumer applications where security is important);

– lack of ultraviolet and infrared emission (of the utmost importance for the conservation of fragile items, in museums for example).

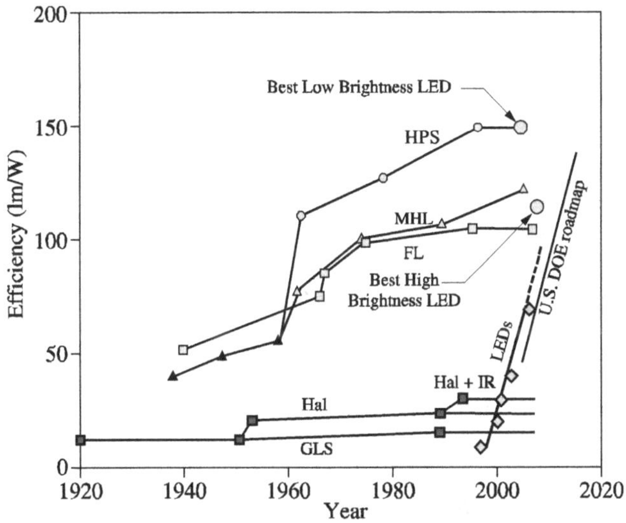

Figure 1.4. *Luminous efficiency of the main technologies for light sources versus time. HPS: High Pressure Sodium (yellow - pink light); MHL: Metal HaLogens; FL: Fluorescent Lights; LED: white LED; GLS: General Lighting System lamps; Hal: Halogens; Hal+IR: Halogens with infrared reflector; DOE Roadmap: goals of the American "white LED" program*

Beyond the energetic issue there also is an economic one. At the moment the lighting market is dominated by three major companies: OSRAM, Philips and General Electric[9]. Within this oligopoly context Europe holds a dominating position today. The technological breakthrough proposed by the uprising of LEDs will probably shake this European domination and reshape the lighting market on a worldwide level.

LEDs are nowadays mainly fabricated in Asia, for standard LEDs[10], and in Japan and the USA for high brightness LEDs. The chart in Figure 1.5

9. These three major companies have a €12 billion accumulated annual income (2005), whereas the total lighting world market reaches €15 billion.

10. Luminous flux less than 50 lm.

compares the needs and production capacities in various regions of the world [EDE 06].

This market is now becoming a commodity market where very few European companies are active. On the other hand, Europe can capitalize its know-how on semiconductors in order to play a leading role on the optimized performance LEDs market. With an annual growth rate of 30%[11], the Chinese LED industry will soon be part of the world leaders (Figure 1.6) [WUL 07].

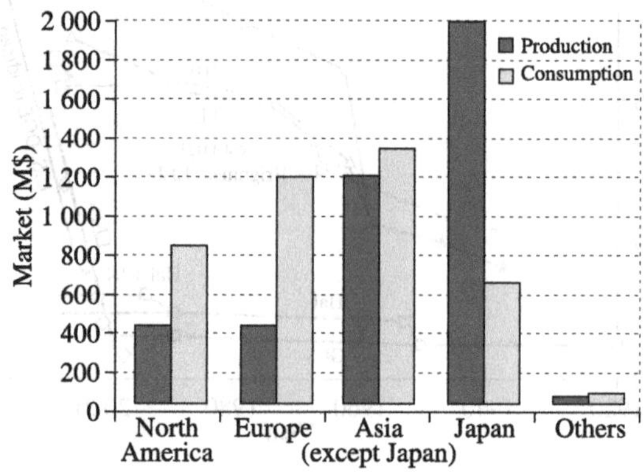

Figure 1.5. *Production capacities and consumption of High Brightness LEDs in the different parts of the world*

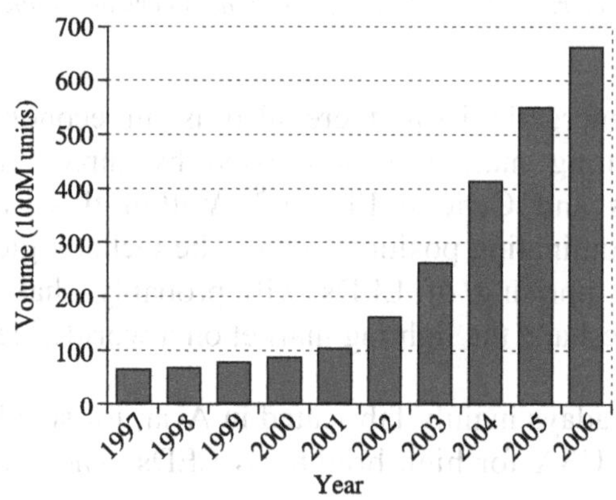

Figure 1.6. *LED production in China (in hundred million units). For comparison, the total number of LEDs produced in 2000 was estimated to be 100 billion units, of which Chinese production corresponded to less than 10%*

11. Annual growth rate based on production volumes for the period 1997-2006.

Today, the overall LED market is worth $8.5 billion. According to Robert Steele [STE 07], director of optoelectronics programs at Strategies Unlimited, High Brightness and Ultra High Brightness LEDs which have a huge market potential estimated at more than $6 billion with an annual growth rate of 8%. If this trend is confirmed and maintained for the next 5 years, this market segment would be worth close to $9 billion in 2009 and more than $10 billion in the following decade (Figure 1.7).

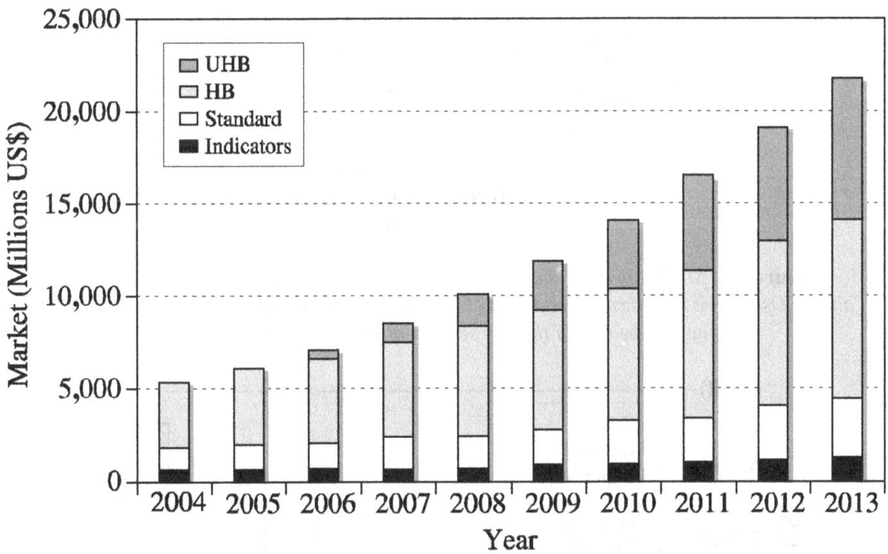

Figure 1.7. *Market shares of the various available types of modern LEDs. Depending on the luminous flux, LEDs are sorted as: indicators: less than 5 lm; standard: 5-50 lm; HB (High Brightness): 50-250 lm; UHB (Ultra High Brightness): more than 250 lm*

If this growth is confirmed, LEDs will become one of the key sector in the microelectronic market before 2020 (Figure 1.8). The development of LEDs therefore becomes vital, not only to save energy, but also to re-boost the semiconductor industry, whose growth rate has been slowed down by the telecommunications sector in the last couple of years.

However, as shown on Figure 1.9 [STE 07], the main applications of High Brightness LEDs are mobile devices, displays and automobiles. Today the market share of LEDs in general lighting is around 6%, likely to reach 10% in 2010. Will this share increase in the coming years? Despite the overall optimism due to the spectacular scientific breakthrough obtained and brought forward by some[12], this question today remains unclear because some technological issues are yet to be solved.

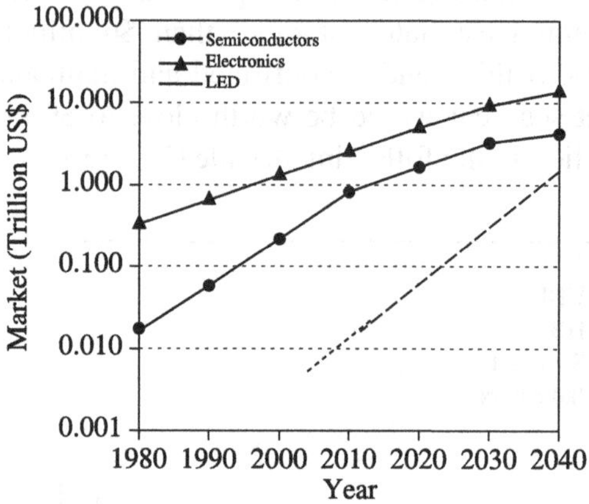

Figure 1.8. *Market predictions for electronics, semiconductors (without LEDs), and LEDs for the next three decades [SZE 02]. If the growth rate of LEDs is confirmed, they will be one of the key sectors of the market before 2020*

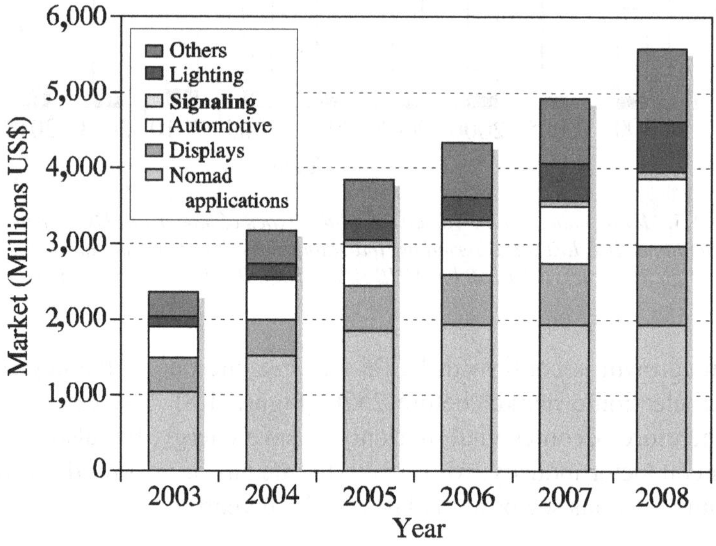

Figure 1.9. *Applications of High and Ultra High Brightness LEDs. General lighting remains behind*

12. Some predictions reach 50% market share of the standard light bulbs in the next couple of years, and even 100% by 2020.

Even if today LEDs are yet to confirm their efficiency in the general lighting systems, they have established themselves in signaling and have begun to penetrate the specialized lighting market. Figure 1.10 summarizes the LED applications as a function of color and luminous flux.

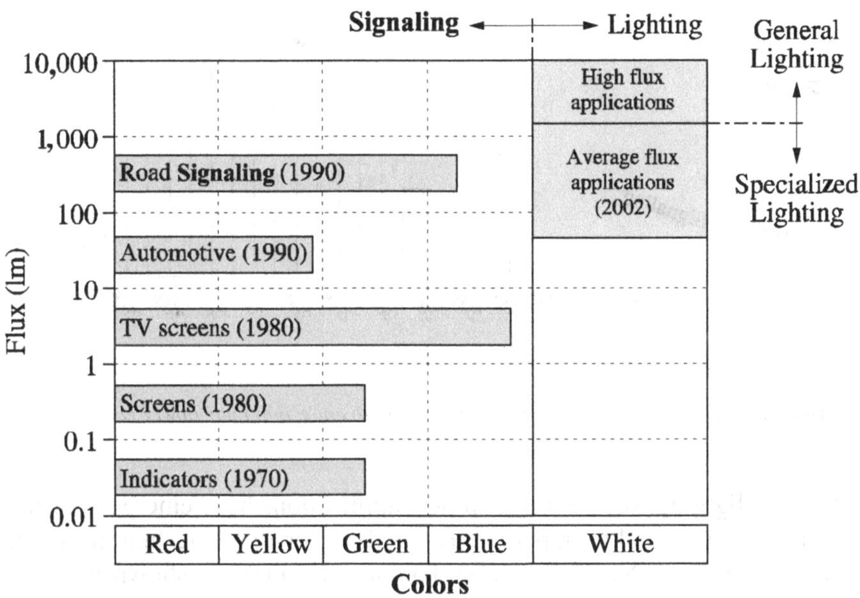

Figure 1.10. *Application domains of LEDs as function of their colors and required flux*

Signaling applications usually require small luminous flux and monochromatic light. In the 1970s, a sole LED was used as an "indicator" (on/off, fault, levels, etc.) on electric and electronic equipments, whereas a few dozen LEDs are nowadays used for a single function on more complex signaling systems (traffic lights for example). Displays require several LEDs of different colors and relatively important flux. Finally, lighting requires white light and powerful luminous flux. Availability of high intensity diodes will no doubt lead to lighting orientated systems, which will function with a limited number of diodes. Knowing this, it is now easier to understand the graph in Figure 1.11 which indicates, according to Steele [STE 07], that LEDs have for a few years reached their "power" zone and now target illumination systems, automotive, back-lighting of large screens and finally lighting.

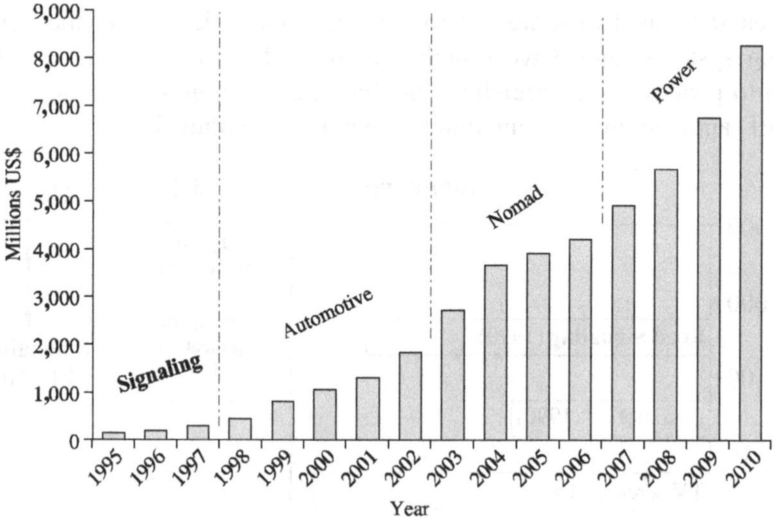

Figure 1.11. *LEDs have now become mature enough to enter into their power zone [STE 07]*

Besides lighting, several original new applications have emerged, such as the illuminations of skyscrapers in Hong Kong or dynamic lighting of the ancient stone "Pont Neuf" bridge in Toulouse (France) as shown in Figure 1.12. LEDs are increasingly entering this promising sector of the "architectural illumination", where they not only enable viewing but also emphasize the artistic features of major buildings.

Figure 1.12. *Left: LED illumination of the Canon tower in Kowloon (Hong Kong) 14 lines with 30 LEDs per line. Each chip consumes 6 W and includes six junctions (2 red, 2 green and 2 blue), for a total power of 12.5 kWh per night [LED 07]; right: arches of the Pont Neuf bridge in Toulouse have been illuminated since November 2007 with a dynamic LED lighting (total power 1 kW)(see color plate section)*

It seems obvious to us that, in order to gain market shares in the general lighting domain, white LEDs need better efficiency and power performances. The next section will discuss the color and efficiency properties of LEDs.

1.3. Principle of operation, color, efficiency, lifetime and quality of LEDs

LEDs are electronic devices that allow electricity to travel in only one direction. They emit light when electric current passes through them.

Like a normal diode, the LED consists of a chip of semiconducting material impregnated, or doped, with impurities to create a p-n junction. At steady state, opposite sign carriers (electrons and holes) cannot penetrate the junction region due to a space-charge field developed.

When the junction of a LED is forward biased, as in other diodes, the current flows easily from the p-side, or anode, to the n-side, or cathode, but not in the reverse direction. When electrons cross the junction from the n- to the p-type material, the electron-hole recombination process produces some photons in a process called electroluminescence. An exposed semiconductor surface can then emit light.

An LED is intrinsically monochromatic and its conversion efficiency depends on the emitting wavelength. Figure 1.13 shows the band diagram of a direct bandgap semiconductor.

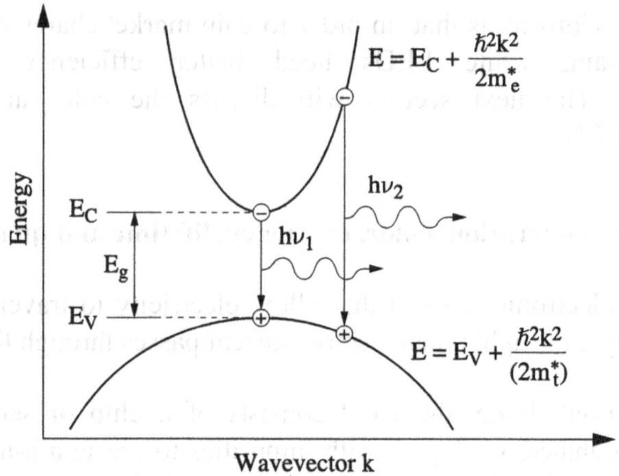

Figure 1.13. *Band diagram of a direct bandgap semiconductor*

During the radiative recombination of an electron located on the minimum of the conduction band with a hole located at the maximum of the valence band (k=0), one photon is emitted with a frequency:

$$h\nu_1 = E_g = E_C - E_V \quad [1.1]$$

where h is Planck constant.

As shown on Figure 1.13, for $k \neq 0$, the photon emitted by the electron hole recombination exhibits a frequency ν_2 slightly different than ν_1. The LED will therefore emit a ray with a central frequency ν_1 and a Full Width at Half Maximum (FWHM) $\Delta\nu_{FWHM}$. This width can be estimated by taking into account its dependency on the density of electrons available in the energy range $E \pm \Delta E$, and also on the density of energy states of the semiconductor within that energy range. For a junction working at a temperature T with a bandgap energy E_g, the electron density is given by Boltzmann's law and is proportional to $\exp(-E/k_BT)$. In the same conditions, the energy level density is proportional to $(E-E_g)^{1/2}$. The theoretical FWHM is therefore given by:

$$\Delta E_{FWHM} = h\Delta\nu_{FWHM} = 1{,}8 k_B T \quad [1.2]$$

whereas the maximum intensity is obtained at the frequency $\nu_{max} \neq \nu_1$:

$$\nu_{\max} = \frac{E_g}{h} + \frac{k_B T}{2h} = \frac{c}{\lambda_{\max}} \qquad [1.3]$$

So, for a junction with E_g=2.5 eV working at 25°C, we obtain $\lambda_{\max} \approx 490$ nm and $\Delta\lambda_{FWHM} \approx 9$ nm. The light can therefore be considered as quasi-monochromatic because in practice, in the light sources' world, there are rays whose width can be higher than 20 nm.

In fact, the first commercialized LED was red. Today, almost all saturated colors are achievable.

Note that the values given by relations [1.2] and [1.3] strongly depend on the junction temperature. The LED color can therefore drastically shift with time if the junction temperature changes (Figure 1.14). When the temperature is increased, the spectrum shifts towards a longer wavelength (for example, amber shifts to red). This shift is in the order of 0.1nm/°C[13].

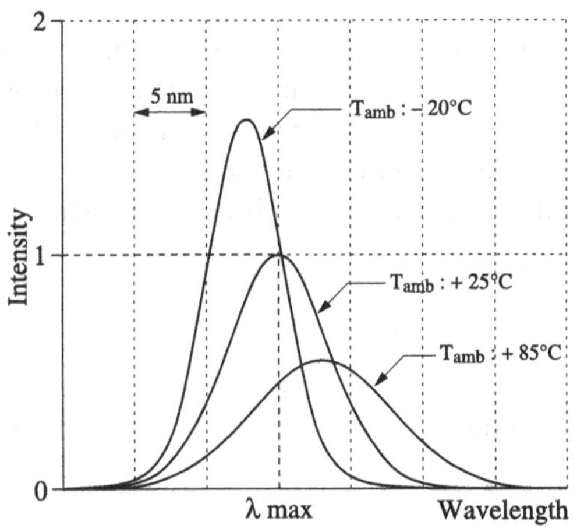

Figure 1.14. *LEDs emit quasi-monochromatic spectra, but the intensity and the spectral width strongly depend on the junction temperature*

13. Variations of the colorimetric characteristics of a LED can cause serious issues for designers. For many applications there are often strict standards regulating the colorimetric characteristics of the equipment. This is even more drastic for automotive and avionic applications where the wavelength and width are strictly normalized.

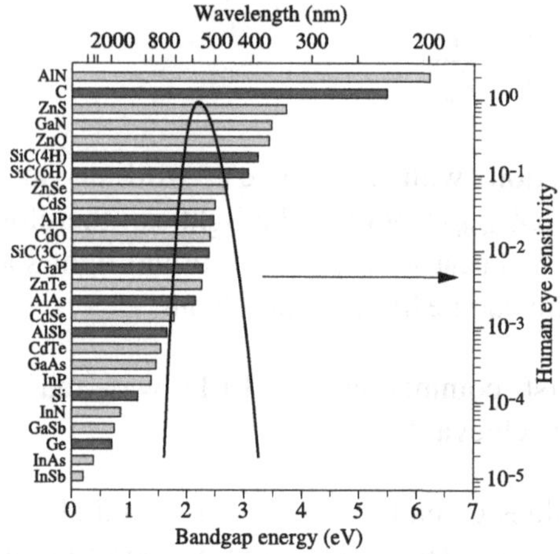

Figure 1.15. *Bandgap energy wavelength for some well-known semiconductors and binary systems. Only direct bandgap semiconductors (light gray bars) are suitable for high intensity LEDs. Indirect bandgap semiconductors (dark bars) are rarely used today. The curve shows the human eye sensitivity $V(\lambda)$ as a function of the energy*

Figure 1.15 shows the gap energy and emitting wavelength for a few well-known semiconductors and binary systems. All wavelengths from infrared to the near-ultraviolet are reachable. For technological reasons only direct bandgap semiconductors can be used for high brightness LEDs. The use of indirect bandgap semiconductors is nowadays still very limited.

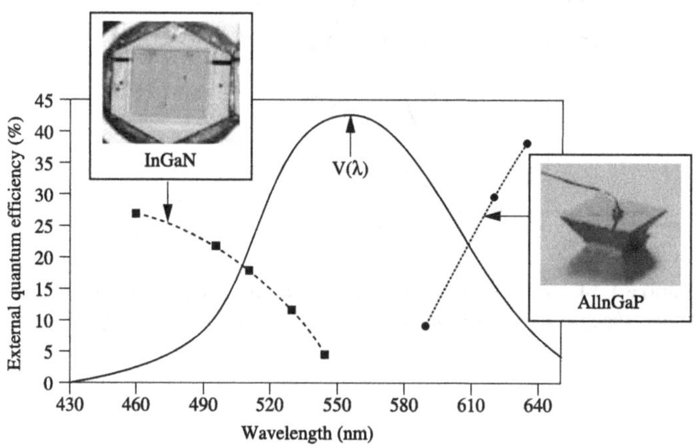

Figure 1.16. *External quantum efficiency of two types of LEDs as function of the emitted wavelength (350 mA, 25°C). The curve $V(\lambda)$ represents the human eye sensitivity*

It can also be noted that the luminous efficiency strongly varies from one color to another (Figure 1.16). The same figure also shows that the efficiency quickly drops at the wavelength close to the maximum human eye sensitivity for today's most used two families of semiconductors (InGaN and AlInGaP).

One of the key issues for LEDs in order to significantly penetrate the general lighting market is to obtain a "high power white LED" with an efficiency better than 100 lm/W, whereas today's most powerful components available on the market show efficiencies of 30-40 lm/W.

The officially announced goal of American researchers is to fabricate, before 2011, white LEDs with efficiencies of 200 lm/W, a lifetime of more than five years with a cost of less than $1/diode. It certainly is a very ambitious project with an uncertain result, but in 2001 the American Congress voted for the S.1166 Act granting $430 million for the achievement of the project [SEN 01]. The European Union, within its technological platform Photonics21, also gave itself in 2005 the goal of reaching a white LED with 150 lm/W by 2030, which seems more realistic. However, the budget allocated to this action is yet to be decided [PHO 06].

1.3.1. *White light production from LEDS: principles and challenges*

There are today three methods to generate white light for a LED:

– combine a diode emitting at short wavelength λ1 with a phosphor emitting at a larger wavelength λ2;

– use a diode, emitting in the near ultraviolet, coupled with one or several phosphors;

– use three diodes (at least) emitting at different visible wavelength which then combine themselves to produce white light.

Each of these methods presents advantages and drawbacks.

The first method is based on the fact that two photons of complementary wavelengths (λ_1 "short" and λ_2 "long") arriving simultaneously on the human eye will cause a white light sensation (Figure 1.17).

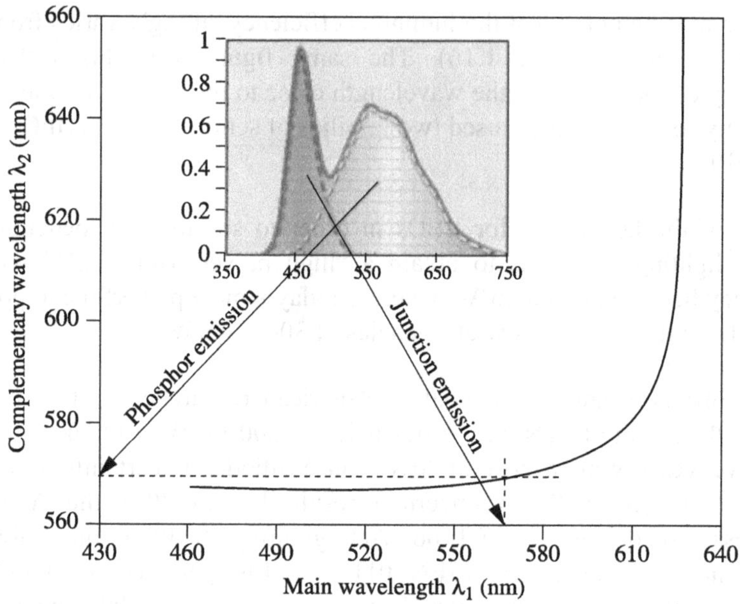

Figure 1.17. *By mixing a short wavelength λ_1 with its complementary λ_2, we obtain white light. Therefore, a blue junction with a yellow phosphor gives white light*

The principal idea is therefore to use a diode generating a short wavelength, covered with a phosphor[14] which absorbs a few short photons to convert them into a longer wavelength. In principle, an endless list of combinations is possible. In practice, several companies (Nishia, CREE, Lumileds, etc.) mass-produce high power white LEDs (now reaching 5W) by using a blue diode (InGaN) combined with a yellow phosphor (YAG-Ce). This combination makes it possible to obtain LEDs with a color temperature in the order of 5,500 K. To produce "warm white" with a color temperature in the order of 3,200 K, an extra layer of phosphor emitting red light is added. However, this significantly reduces the luminous efficiency of the LED.

In the short term, this technology will probably dominate the white LED market.

14. We could also consider using another LED emitting yellow light since two photons of complementary wavelength produce white light. However, with two monochromatic lights, the color rendering would not be very good: this problem is solved by phosphor which exhibits a large spectrum.

However, it presents two main drawbacks:

– the diode has a halo effect which can be troublesome for some applications. This is due to the generation of directional light from the diode, whereas phosphor has a "Lambertian" behavior (diffuse emission);

– yellow phosphor does not easily absorb blue photons because of the rare earth used as dopants. The solution of this problem seems complex and would require a considerable amount of fundamental research.

Reaching an efficiency of 200 lm/W for white light production with this technology coupled with existing phosphors would require a diode with a total conversion efficiency of 60%[15]. The best efficiency reported to date is about 25-30% at a wavelength of 450 nm.

The second method consists of using a short wavelength emitting diode (near ultraviolet) coupled with one or several phosphors which convert the ultraviolet light into visible light. The same method is used in the case of fluorescent lights. The main advantage is that it produces very high quality white light (good color rendering).

Applying this technology to reach 200 lm/W requires a UV diode with a total efficiency of 70%. Diodes generating photons at 400 nm do exist but their efficiencies hardly reach 21%. To increase it to 60-70% is a formidable challenge but seems unrealistic considering today's state of the art, even if there is a very large amount of research activity on that topic. The development of efficient UV diodes with long lifetimes and reasonable power is an important issue not only for lighting but also for other applications such as dermatology, surface treatment, decontamination, etc. If this technology matures, and in particular if reliable good quality phosphors become available, this method could well become the leader on white light LEDs.

Finally, *the third method* consists of using three LEDs, one for each fundamental color (red, green and blue). The synthesis of these three colors leads to white light production whose color temperature depends on the ratio of each color. More than three sources are usually used today. This makes it possible to obtain more color variations or, to be more accurate, variations

15. The "total conversion efficiency" is the ratio between the emitted optical power to the electrical power supplied to the component.

on the desired color. It is therefore possible to add cyan, amber or red-orange, the main goal being to improve the color rendering.

Considering that the dimensions of a diode are small, it is also possible to imagine multiple chip white light sources embedded into a material diffusing the light in order to improve the source homogenity. This method prevents the use of phosphors and enables a good color control.

However, this technology has a major drawback: it requires an individual control of each diode from the few tens or hundreds included in the source but also a different voltage supply for each chip, and therefore an increase of the overall cost of the source. Finally, compared to a classical light source of comparable luminous flux, the overall surface is much larger. Applying this solution to reach 200 lm/W implies the use of LEDs with efficiencies of 30-40% for each color. More details on white light spectra emitted by LEDs can be found in section 7.3.

Table 1.1 summarizes the advantages and drawbacks of the three methods.

	Advantages	**Drawbacks**
Method 1	• Good luminous efficiency • Proven technology • Various color temperatures (warm or cool shades) • Good Color Rendering Index (CRI) of 70-90	• Limited color variations • Handpicking of the components at the end of the production chain to gather similar color chips • Colored "halo" effect
Method 2	• Good color uniformity • Large white range	• Reduced luminous efficiency • Small output power • UV packaging
Method 3	• Dynamic color control • Possibility to produce millions of colors	• Complex electronic control • Temperature sensitive colors • Requires flux homogenity • Small CRI

Table 1.1. *Methods for white light production using LEDs*

1.3.2. *Lifetime*

A few years ago, the main manufacturers claimed that their products had lifetimes of better than 100,000 hours[16]. However, their definition of "lifetime" was not given: average, median, economic, for the bare junction or for a given lighting system?

The accurate definition of the lifetime for a LED is yet to be normalized. It remains somehow vague but efforts to bring a standard together are underway and more rigorous, realistic tests are undertaken at an industrial level.

It is generally accepted that the lifetime of a LED is defined as follows: it is the time after which, in a given sample of LEDs, a ratio "B%" of this sample emits a luminous flux "L%" lower than the initial flux. B50 L70 therefore means that 50% of the LEDs in this sample will emit only 70% of the nominal flux after a time "t". It roughly corresponds to the lifetime definition used for other light sources, even if the test conditions are not yet specified. It is to be noted that for LEDs, the initial luminous flux is generally measured after 24 working hours, whereas for classical light sources the nominal power is measured after 100 hours of monitored operation[17].

The general acceptance of this more realistic definition, with respect to the lighting world standards, explains why the values originally announced have been reduced.

The lifetime of a LED strongly depends on the operating conditions, the main factors being:

- the junction temperature;
- the applied current.

The junction temperature plays a major role in the reliability of LEDs. The announced values are generally given for temperatures of around 20-25°C. It is commonly agreed that the lifetime is divided by a factor of two as the junction temperature is increased by 10°C. By placing the LED in a lighting

16. For some systems, however, probably badly designed, lifetime hardly reached 5,000 hours.
17. During the initial burn-in, the light bulbs remain switched on for 11 consecutive hours, then off for one hour and the cycle restarts.

system, the lifetime can therefore be drastically reduced without thermal management. The graph in Figure 1.18 clearly shows this dependency.

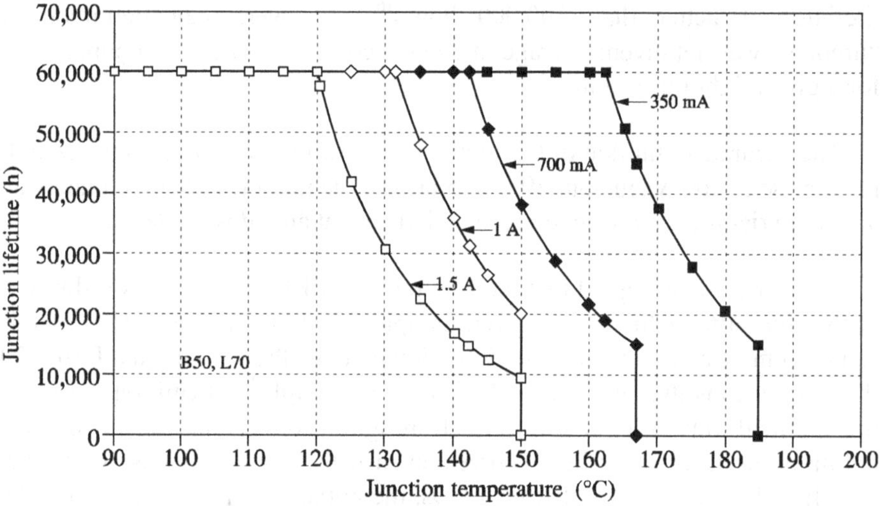

Figure 1.18. *Dependency of the lifetime of a LED as a function of the bias current and the junction temperature (B50, L70 mean that 50% of the LEDs from the tested sample emit 70% of the initial flux)*

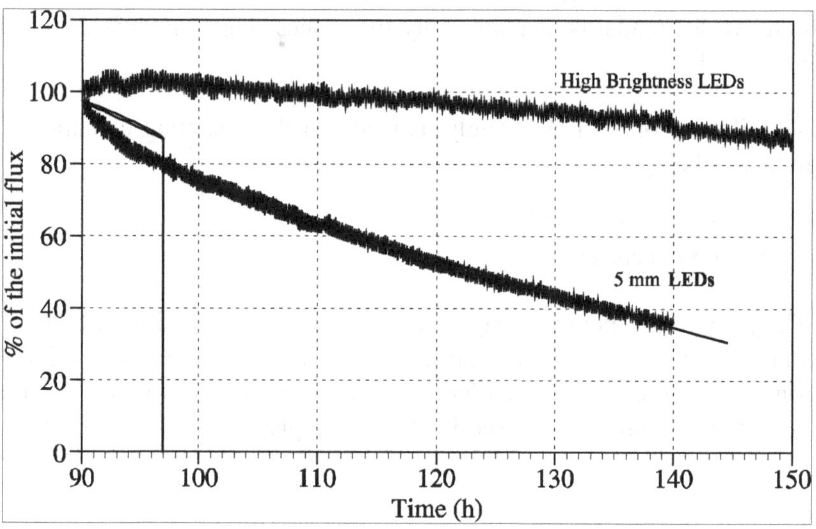

Figure 1.19. *Reduction of the flux for 5 mm and High Brightness LEDs compared to incandescence lights*

It is important to emphasize the huge progress achieved for keeping the luminous flux, especially for high brightness LEDs. Figure 1.19 shows the reduction of the flux as a function of time for the main LED technologies (5 mm and High Brightness).

Finally, the aging process of LEDs is rather erratic, the color shifting with time. Some improvements in that field are therefore also required for the LEDs to dominate the lighting market.

1.3.3. *Quality of LEDs*

This is often a missing point in the discussions on the advantages of LEDs. However, it is of the utmost importance for system reliability and market penetration.

Figure 1.20 illustrates this issue: keeping the flux from identical LEDs from various providers show great disparities.

This situation can be a major concern because the low or bad quality LEDs existing today on the market give bad publicity to the system designers and the final users for the overall LED products.

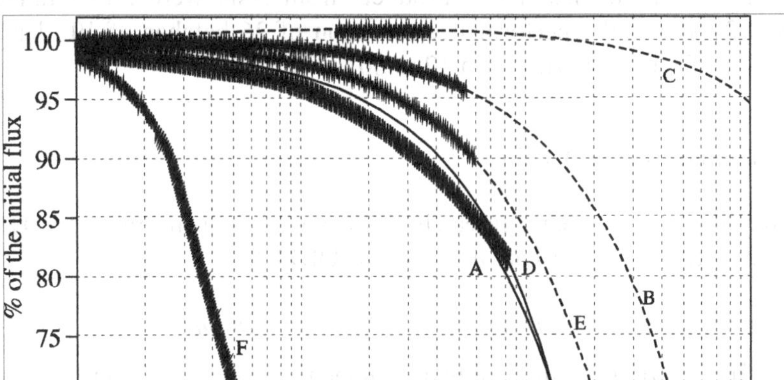

Figure 1.20. *Reduction of the luminous flux of identical high brightness white LEDs from various providers (letters A to F). Aging is carried out in the same conditions (current, junction temperature)*

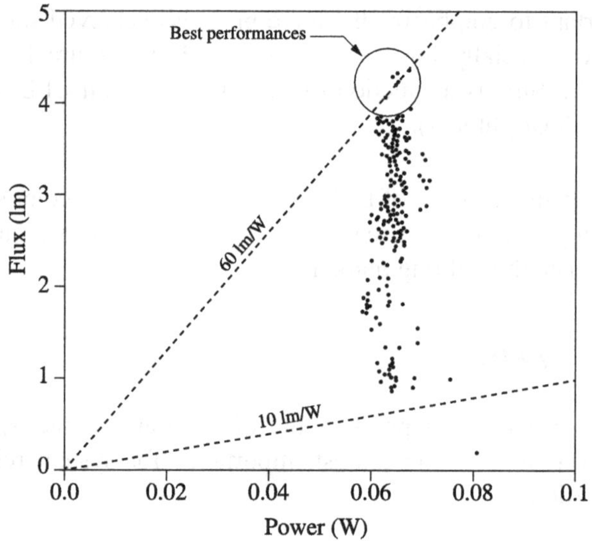

Figure 1.21. *Variation of the performances for 260 white LEDs (coming from 26 batches) of 5 mm operating at 20mA (LEDs made in Asia)*

On top of this, some reproducibility concerns also arise from large series production. A recent study led by Evans Mills [MIL 07] showed that the performances of identical batch products from Asia were rather random. Figure 1.21 shows the performance variation of 26 batches of 10 identical LEDs. The efficiency varies from 12 to 60 lm/W.

1.4. Challenges facing LEDs

Today there is an intense research effort worldwide concerning LEDs, the various directions of interest are listed below [OID 02].

Fundamental research on materials and light generation

The first important issue is a better understanding of the light generation mechanism in semiconductors. In order to achieve this, new diagnosis techniques specifically dedicated to LEDs and more sophisticated mathematical models than those that currently exist are needed. In the meantime, a better understanding of the p-doping defects and of the incorporation of high concentration of indium and aluminum in green InGaN and AlGaInN diodes would improve the performances of green LEDs. More studies on the carrier confinement in phosphor-based LEDs seem to be the key for the fabrication of more efficient red LEDs. On a material level, indium, one of the key elements for many of the existing LEDs, is a rare material: 61st rank in terms of quantity found on Earth (0.24 ppm in weight). Although a LED junction only uses a very small amount of this element, it has so far been impossible to recycle after use. Today's reserves worldwide are estimated to be 5,600 tons and should be sufficient[18] for the years to come [JOR 05], but they will not be unlimited[19].

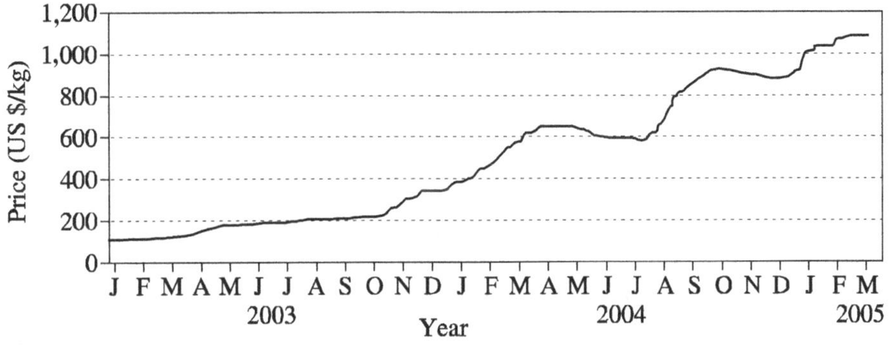

Figure 1.22. *Indium price variations (99.97% pure)*

Moreover the price of this metal has been constantly increasing for the last few years (Figure 1.22). The use of alternative materials like pure silicon (Si) or zinc oxide (ZnO) seems promising even if the technology is still in its infancy.

Improvement of the Internal Quantum Efficiency (IQE)

This is one the major challenges, especially at a time where most of the other technological factors limiting the global efficiency of a LED are being mastered. Figure 1.22 shows the theoretical IQE and the IQE value for a good quality blue LED. The gap is quite large, especially at high currents (and therefore at high output powers). Today's goal is to reach an IQE value of the order of 90%, hopefully in the next decade.

18. The total world consumption was about 400 tons in 2004.

19. Note that 45% of today's indium consumption concerns ITO (Indium Tin Oxide) fabrication for transparent electrodes and an additional 30% is used for electronic component production.

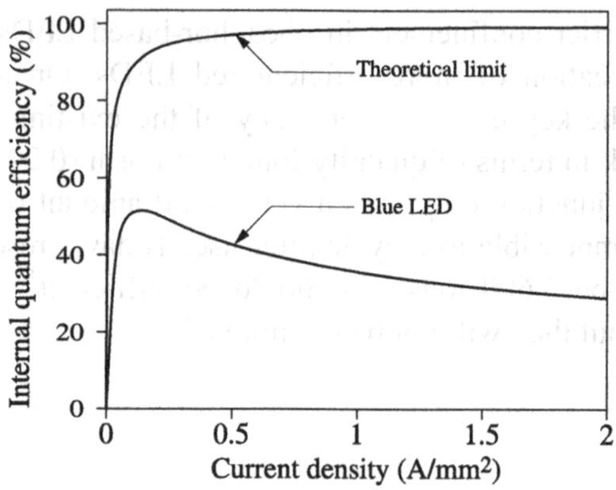

Figure 1.23. *Internal quantum efficiency of a LED as a function of the current density*

Materials for substrates

The substrates used today for the most promising GaN LED fabrication are made of sapphire, SiC or GaN. Each of these materials has its advantages and disadvantages. The main issue is to fabricate a large area with very little defects with these substrates. Solving this issue has to be the main priority in this domain. It would also be best to reduce the cost of fabrication to less than €70/cm^2 (for UV, green and blue diodes).

Epitaxial reactor technology diode fabrication

The reactors used for mass production of the GaN LEDs are at the moment modified GaAs-based diode fabrication equipment. Their reliability is excellent for GaAs diodes but is very poor for modern GaN diodes. Diodes produced on this type of equipment systematically exhibit efficiencies lower than the prototypes by a factor or 1.5 to 2. The characteristics of a given batch produced on this equipment also show great discrepancies. The typical example of this is the hand picking of the same color components for mass produced high brightness LEDs. The optimization of the reactor and LED fabrication require a better understanding of the process chemistry and nitride growth mechanisms.

Phosphors

The issue of poor blue light absorption by yellow phosphors has already been mentioned. Firstly, it is important to develop phosphors able to absorb and convert photons efficiently with a wavelength ranging from 370 to 470 nm. Secondly, it would be useful to study other combinations of complementary colors and to develop dedicated phosphors. Finally, as is the case already for fluorescents lamps, the development of "quantum-splitting phosphors"[20] could be a breakthrough for LEDs.

Light extraction and packaging

The improvement of the optical characteristics of substrates, transparent contacts, reflecting layers, etc., is also an important research direction in the near future. Improving the junction geometry to increase the light extraction efficiency or using more complex structures such as photonic crystals are medium-term challenges. Finally it is important to keep in mind that the junction is only a part of the final product and therefore that the packaging of the diode is of the utmost importance for its efficacy and reliability. Some technological challenges have to be addressed in this domain. For example, in the case of a High Brightness diode, it would be necessary to use a transparent and easily manageable material with a refractive index above 1.6 and a transmittance of 80% from 440 to 650 nm that is also able to block short wavelength photons (λ < 440 nm). This material should keep its characteristics for at least 50,000 hours while facing temperatures of about 150°C. Finally, the water concentration in this material should stay in the order of a few parts per million (ppm) and its mechanical characteristics should be compatible with the application planned.

Reflectors and electronic

In order for LEDs to penetrate the lighting market, specifically designed reflectors are needed, which take into account the strong directivity of the light, the small dimensions of the light source and its high sensitivity to temperature, humidity and other corroding factors potentially present in its environment. Finally, considering the production of a large amount of light necessary for lighting and therefore the large number of LEDs required for this purpose, a sophisticated electronic command is needed. As mentioned before, the color of a LED is sensitive both to the voltage and bias current. A small variation of one of this value would shift the emitted color. The behavior of each diode also changes with time (flux reduction, color shifting) but also

20. *Quantum-splitting phosphors*: technology which makes it possible to obtain from one initial UV photon of energy E two or three visible photons of energy E1, E2, E3 (E>E1+E2+E3). The idle case, but not demonstrated to date, is the generation of one blue, one red and one green photon to produce white light. These phosphors tend to be fairly expensive and also imply recycling after use. This technology will mainly be useful for fluorescent lamps, with or without mercury, and to UV LEDs.

varies from one chip to another due to the fabrication tolerances (see above). It becomes necessary to monitor each diode individually.

1.5. Bibliography

[EDE 06] EDEN, "La technologie des diodes électroluminescentes blanches pour l'éclairage", *Synthèse de la journée nationale d'études sur la révolution de l'éclairage par les diodes électroluminescentes*, Sophia Antipolis, p. 26-29, 29 June 2006.

[HAI 02] HAITZ R., KISH F., TSAO J., NELSON J., "Another semiconductor revolution: this time it's lighting", *Compound Semiconductor Magazine*, March 2002.

[HOL 62] HOLONYAK N., BEVACQUA S.F., "Coherent (visible) light emission from $Ga(As_{1-X}P_x)$ junctions", *Applied Physics Letter*, vol. 1, n°4, p. 82-83, 1962.

[JOR 05] JORGENSON J.D., GEORGE M.W., "Mineral Commodity Profiles – Indium", *Open-File Report 2004-1300*, U.S. Geological Survey, Reston, Virginie, US, 2005.

[LED 07] LED MAGAZINE, "Laservision LED lamps illuminate Canon building in HK", http://www.ledsmagazine.com/news/4/3/6, 1 March 2007.

[LOS 29] LOSEV O.V., *Soviet patent 12191* (1929).

[MIL 07] MILLS E., JACOBSON A., "The Need for Independent Quality and Performance Testing of Emerging Off-grid White-LED Illumination Systems for Developing Countries",*Lumina Project, Technical Report* #1, 2 August 2007.

[OID 02] OIDA, "Light Emitting Diodes (LEDs) for General Illumination", *White LED Road Map* (updated), OIDA, 2002.

[PHO 06] PHOTONICS 21, "Towards a Bright Future for Europe – Strategic Research Agenda in photonics", *Photonics21 European Union Joint Technology Platform Roadmap*, April 2006.

[ROU 07] ROUND H.J., "Note on carborandum", *Electrical World,* vol. 49, p. 309, 1907.

[SEN 01] THE US SENATE, "To establish the next generation lighting initiative at the Department of Energy, and for other purposes", *US Senate Bill S.1166*, 11 August 2001.

[STE 07] STEELE R., Opening keynote talk, *Strategies in Light*, San Jose, 20-23 February 2007.

[SZE 02] SZE S.M., "Microelectronics Technology: Challenges in the 21st Century", in S. Luryi, J. Xu and A. Zaslavsky (Eds) *Future Trends in Microelectronics: The Nano Millennium*, John Wiley & Sons, New York, 2002.

[WUL 07] WU L., "National SSL program in China", *Proceedings of the 11th International Symposium on the Science and Technology of Light Sources*, Shanghai, China, 20-24 May 2007, p. 479-485.

Chapter 2

Substrates for III-Nitride-based Electroluminescent Diodes

2.1. Introduction

Since the first blue InGaN/GaN electroluminescent diodes with optical output powers surpassing the milliwatt range [NAK 94] were demonstrated 15 years ago, the attention given to III-N semiconductors has exploded.

These results were outstanding considering the fact that the material was showing very high crystal defect densities (~ 10^9-10^{10} dislocations/cm^2), more than four orders of magnitude higher than those in III-As or III-P LEDs.

The defects seen in nitride LEDs mainly result from stresses due to their epitaxial growth on foreign substrates.

These substrates exhibit differences in lattice parameter, chemical composition and coefficient of thermal expansion (CTE) with GaN. The differences in CTE are responsible for cracks due to thermal stress during the temperature cycle while growing.

Chapter written by Philippe DE MIERRY.

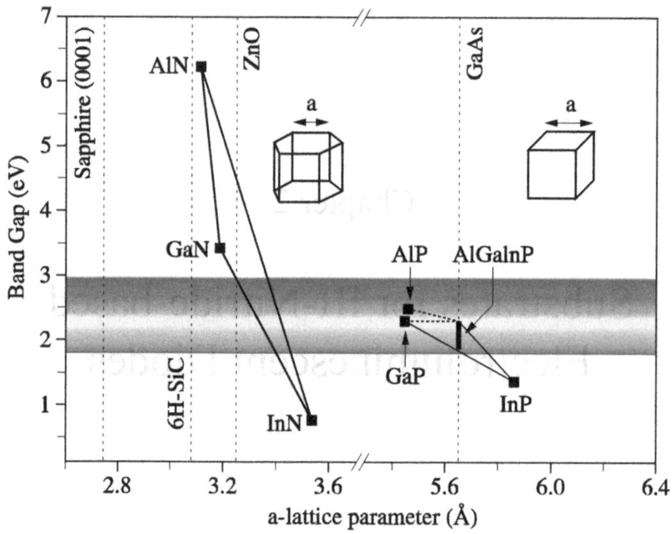

Figure 2.1. *Bandgap of nitrides (Al,Ga,In)N and phosphides (Al,Ga,In)P used for visible LEDs, as a function of their lattice parameter (see color plate section)*

Figure 2.1 shows the lattice parameters as well as the forbidden energy band of the main semiconductors used for the fabrication of visible LEDs. Red-orange LEDs are fabricated from the AlGaInP cubic alloy for which the epitaxial growth can be lattice-matched with GaAs. With higher Al content the light emission moves towards blue but the efficiency quickly falls below 590 nm (>2.1 eV) since the forbidden energy band of the alloy becomes indirect [SCH 99]. However the (Al, In, Ga)N hexagonal system, with a direct gap, potentially achieves a large spectral band from infrared to visible and ultraviolet. No substrates perfectly match this alloy for epitaxial growth. The lattice parameter of ZnO is close to GaN but it is chemically unstable under the normal growth conditions of GaN [LEE 07]. A large number of other substrates have been used for the heteroepitaxial growth of GaN, among them $LiGaO_2$, GaAs [LIU 02], Si [FEL 01] and the spinel $MgAl_2O_4$ [TIN 05].

Without bulk GaN substrates, Al_2O_3 sapphire and polytype 6H-SiC carbide were and still are the substrates used in the nitride LED industry.

The lack of proper substrates for nitrides is a serious impediment; it limits the performances and lifetime of optoelectronic components. (In, Ga)N quantum well based LEDs are however less sensitive to crystal defects than those with arsenide and phosphide [LES 95]. Fluctuations of potential in (In,Ga)N well provoked by composition changes (due to the co-existence of rich and poor indium region) explain the specific behavior of nitrides. The injected carriers tend to localize at the minimum potentials where they recombine radiatively.

The beneficial role of indium in the quantum wells of visible (In,Ga)N LEDs is illustrated in Figure 2.2 where the internal quantum efficiencies of III-N LEDs are plotted as a function of the dislocation density. The figure clearly shows that the efficiency of (In,Ga)N LEDs is not as affected as GaN LEDs.

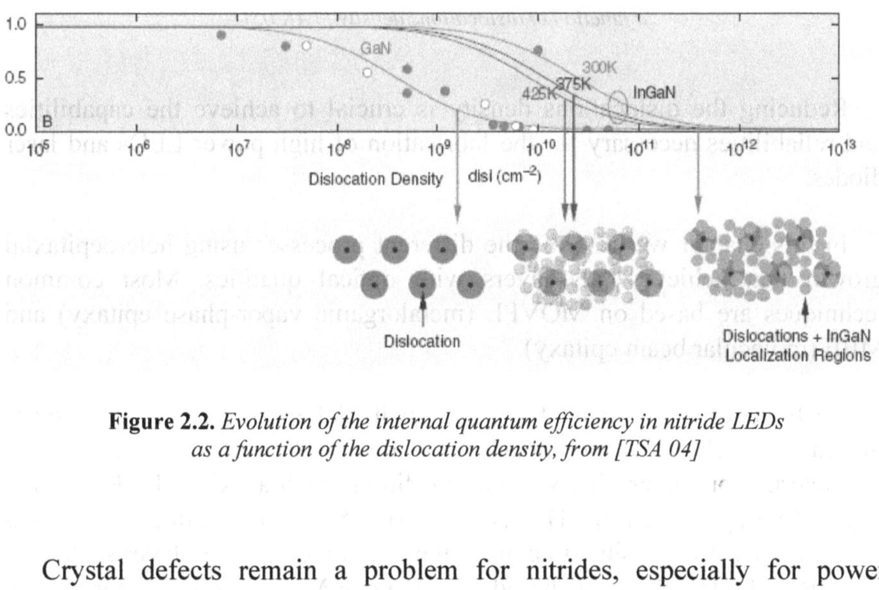

Figure 2.2. *Evolution of the internal quantum efficiency in nitride LEDs as a function of the dislocation density, from [TSA 04]*

Crystal defects remain a problem for nitrides, especially for power devices where high currents are used, and for UV (Al,Ga)N LEDs. It is also proven that the lifetime of (In,Ga)N laser diodes decreases with increasing density of dislocations. Figure 2.3 shows that this density must remain around $3x10^6/cm^2$ to obtain a lifetime of 10^4 (420 days) hours compatible with the commercialization of lasers.

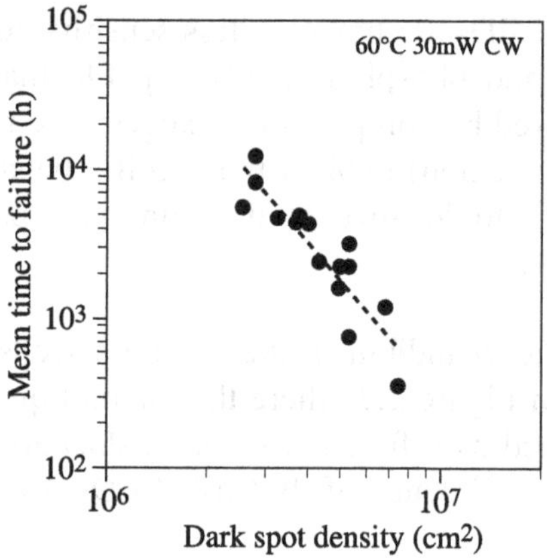

Figure 2.3. *Operating lifetime of nitride laser diodes as a function of dislocations density [TAK 03]*

Reducing the dislocations density is crucial to achieve the capabilities and reliabilities necessary for the fabrication of high power LEDs and laser diodes.

In this chapter we describe the different processes using heteroepitaxial growth that achieve GaN layers with optical qualities. Most common techniques are based on MOVPE (metalorganic vapor-phase epitaxy) and MBE (molecular beam epitaxy).

MBE is readily suitable for the growth of heterostructures with sharp interfaces such as superlattices and quantum wells. Using *in situ* characterization under high vacuum conditions such as RHEED (Reflection High Energy Electron Diffraction) or STM (Scanning Tunneling Microscopy), it is possible to control the growth at the monolayer scale (2.5 Å) [GRA 98], [VEZ 04]. It should be noted that MOVPE uses higher growth temperatures than MBE. It is therefore possible to use a solid phase epitaxy during the first nucleation phases, which is a key step when the substrate is dramatically different from the epitaxially grown material. It also has a higher growth rate and for this reason thick layers of GaN for opto-electronics devices are always achieved using MOVPE.

All the problems resulting from heteroepitaxial growth could be solved using homoepitxial growth of components from bulk GaN and AlN substrates. However, these substrates are just emerging and still represent a major challenge for the industry. This subject is discussed at the end of this chapter.

2.2. Crystal structure and epitaxial relation with 6H-SiC and Al_2O_3

III-N semiconductors are formed with an element from the third (III) column (In, Ga, Al) and atomic nitrogen. Their basic lattice structure is the diamond lattice with sp^3 bonding, which crystallizes either in the zincblende structure (for the cubic system) or the wurtzite structure (for the hexagonal system).

The wurtzite structure is the most stable due to the ionic nature of the metal-N bond.

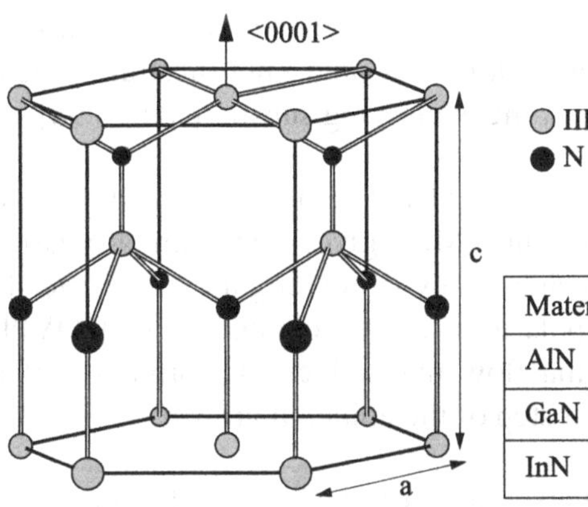

Material	a (Å)	c (Å)
AlN	3.112	4.982
GaN	3.189	5.185
InN	3.538	5.703

Figure 2.4. *Wurtzite structure of (Al,Ga,In)N*

Because the wurtzite structure is polar along the c [0001] axis there is an intrinsic polarization, also called spontaneous polarization.

Nitride polarity is defined by the orientation of the III-N bonding with respect to the sample surface (see Figure 2.5). We distinguish surfaces with Ga polarity or N polarity.

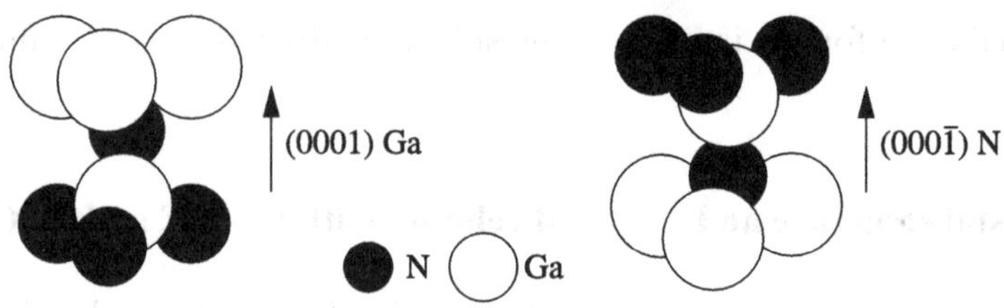

Figure 2.5. *Orientation of atomic bonding for Ga and N polarity in GaN crystal*

The two most common substrates for the epitaxial growth of GaN are Al_2O_3 sapphire with a trigonal crystal lattice and 6H-SiC silicon carbide with a hexagonal lattice. Both substrates usually have an [0001] oriented epitaxial surface and the growth of nitride follows this orientation.

The N-face or Ga-face polarity will then depend on the substrate and the method used for growth. With polar 6H-SiC and ZnO substrates, the polarity of GaN replicates the polarity of the underlying substrate [HEL 98].

The Al_2O_3 substrate is, on the other hand, non-polar. The direct growth of GaN on Al_2O_3, using nitrogen source MBE gives GaN with N-polarity [CHO 03], whereas Ga polarity is achieved by depositing an AlN buffer layer on sapphire prior to growth. Ga polarity is always obtained with MOVPE as the high temperature and ammonia flow rate induce the formation of an AlN monolayer on Al_2O_3 at the first step of the epitaxial growth.

The epitaxy of GaN layers may also be achieved with the C axis disoriented with the growth direction. The nitrides are then called semi-polar or non-polar. With substrates such as 6H-SiC, which are symmetric to GaN, the crystalline orientation of the GaN follows the substrate precisely. With substrates such as Al_2O_3, the epitaxy can follow different orientations depending on the crystal plan of Al_2O_3 (see Figure 2.6). For a plane with a C (0001) orientation or A (11-20) orientation, the GaN film grows along the C polar direction. With R (1-102) orientation sapphire, GaN grows in the non-polar direction A <11-20>, the C axis being parallel to the surface plane. Finally M(10-10) sapphire achieves <11-22> orientation GaN, called semi-polar, with the C axis forming a 32° angle with the surface plane.

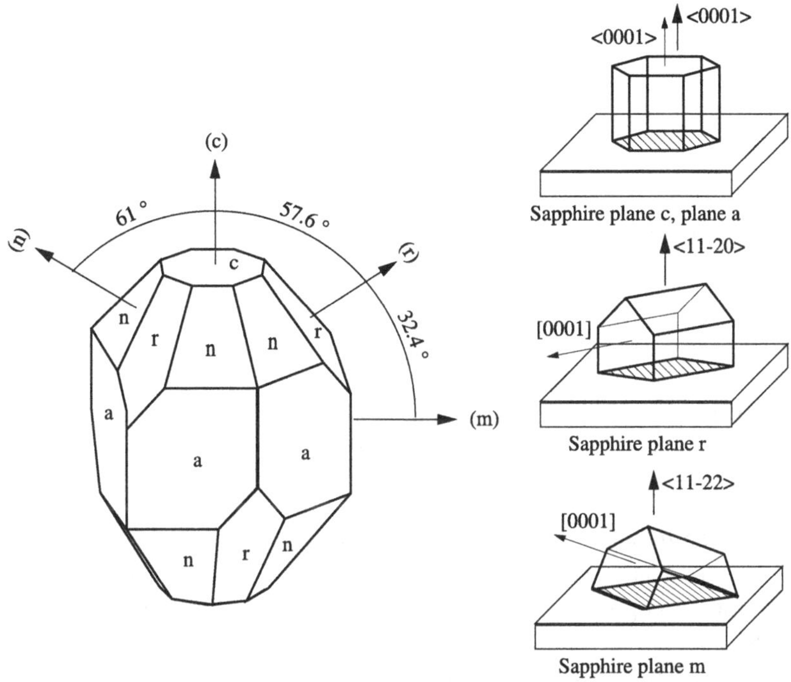

Figure 2.6. *Epitaxial growth direction of GaN in respect to the different crystal plans of* Al_2O_3

Non-polar nitrides do not exhibit a spontaneous or piezoelectric polarization in the growth direction, which eliminates the quantum confined Stark effect in the active structure of LEDs, and thus increases the radiative efficiency. However, the crystal quality of non-polar or semi-polar nitrides is far from those of polar nitrides, mainly because of stacking faults [GUE 07].

The rest of the chapter deals with (0001)-oriented polar nitrides, on which all optoelectronic devices have been based so far.

Table 2.1 reports some important properties of the two main substrates, 6H-SiC and Al_2O_3 in comparison to GaN.

Properties	Al_2O_3 (0001)	6H-SiC (0001)	GaN (0001)
Lattice parameters (Å)	a= 4.758 c=12.98	a=3.08 c=15.117	a= 3.189 c=5.185
CTE (x10^{-6} K^{-1})	7.5	4.46	5.59
σ_{th} (W/m/K)	41.2	490	210
Lattice mismatch with GaN Δa/a (%)	-13.8	-3.4	0

Table 2.1. *Comparison of some substrate properties with GaN. CTE = coefficient of thermal expansion; σ_{th} = thermal conductivity*

The lattice mismatch between sapphire and GaN is higher than with SiC, which is a disadvantage for heteroepitaxy. However, with advanced technologies it is possible to achieve similar crystal quality regardless of the substrate used. The main advantages of SiC are its very high thermal conductivity, which achieves better thermal dissipation of LEDs, as well as its very good electrical conductivity with which a back side metallic contact in the LEDs is possible. Sapphire is, on the other hand, thermally and electrically insulating.

There is a difference in the coefficient of thermal expansion (CTE) between SiC and GaN, SiCs being smaller, which causes tensile stress during cooling from the growth temperature (~1,000°C) down to ambient temperature. The GaN layer is then under tension and some cracks may appear for thicknesses above 5 µm [LAH 00]. On the other hand, the CTE of sapphire is smaller than that of GaN. In this case, during cooling, the GaN film is under compression, which may cause cracks to appear at the GaN/sapphire interface and propagate within the GaN [HIR 93].

Figure 2.7 shows the trigonal lattice and the (0001) epitaxial plane of sapphire.

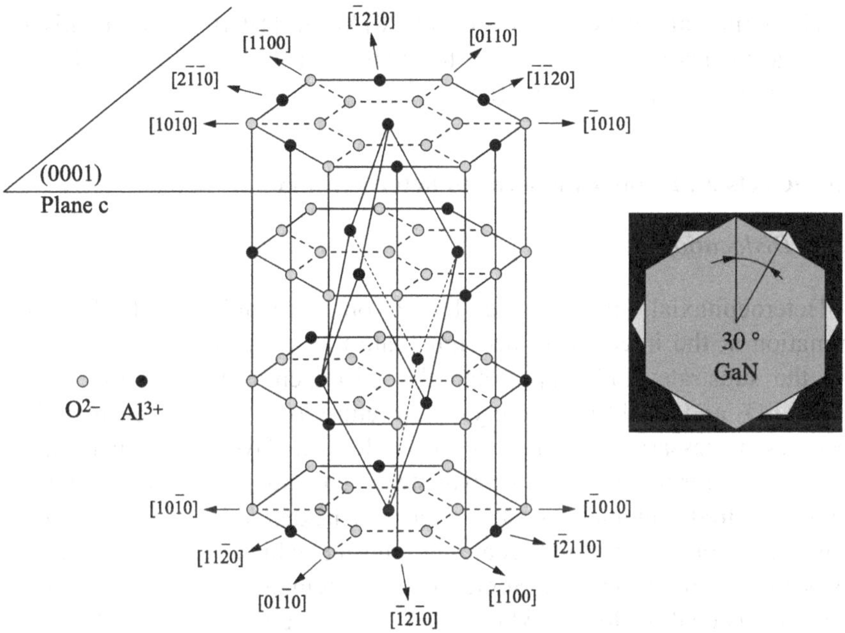

Figure 2.7. *Trigonal lattice and (0001) epitaxial plan of sapphire. The in-plane GaN lattice is rotated by 30° relative to sapphire in order to reduce the parameter mismatch. In this case $\Delta a/a$ = -13.8%*

The lattice parameter of (Al_2O_3) in the (0001) plane is 4.758 Å, whereas for GaN it is 3.189 Å. Without the lattice rotation, the parameter mismatch would be:

$$\frac{\Delta a}{a} = \frac{a(Al_2O_3) - a(GaN)}{a(GaN)} = +50\% \qquad [2.1]$$

Because there is a natural tendency to lower this mismatch, the GaN lattice is rotated by 30° with respect to the Al_2O_3 lattice. In this case, the mismatch is then:

$$\frac{\Delta a}{a} = \frac{a(Al_2O_3)/\sqrt{2} - a(GaN)}{a(GaN)} = -13.8\% \qquad [2.2]$$

The epitaxial relationships are [11-20]Al_2O_3//[10-10]GaN. In this case cleavage planes are very difficult to obtain and ionic etching methods are required to create the laser cavities.

2.3. Defects and constraints due to heteroepitaxy

2.3.1. *Dislocations*

Heteroepitaxial models are based on mechanisms of dislocation formation at the interface to accommodate the mismatch between the film and the substrate [FIT 91]. When the mismatch is too big (as it is for GaN/Al_2O_3 and GaN/SiC) the crystal structure loses its coherence so that it becomes necessary to introduce a nucleation layer, deposited at low temperature during the initial growth stage. This nucleation layer is made of a few hundred nanometers-thick cubic and hexagonal grains. In a second step the grains grow and coalesce, during which a great number of dislocations are generated. Some of them remain near the interface by forming dislocation loops, while others grow vertically along the growth direction [MAT 01]. The majority of defects within the nitrides are threading dislocations resulting from the grain coalescence in the nucleation layer. Figure 2.8 is a transmission electron microscopy (TEM) image that shows a GaN film grown on Al_20_3. We can see the dislocation loops near the interface layer as well as the threading dislocations which propagate through to the top surface.

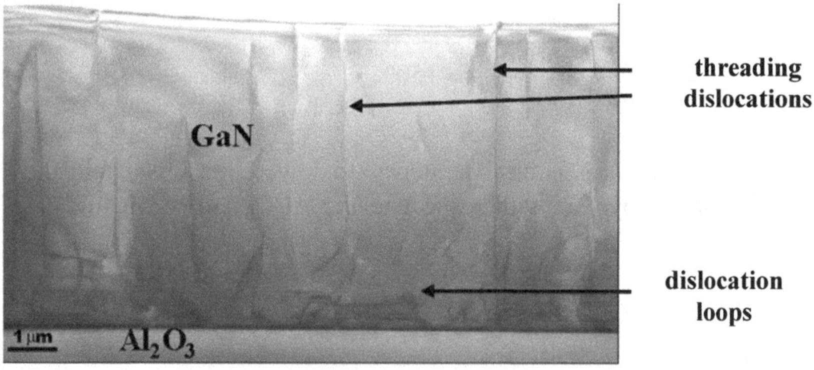

Figure 2.8. *TEM image in cross-section of GaN/Al_2O_3 [P. Vennéguès, CRHEA]*

This type of growth is called columnar growth because of the large coherence length along the (0001) axis and the reduced length in the plane (see Figure 2.9).

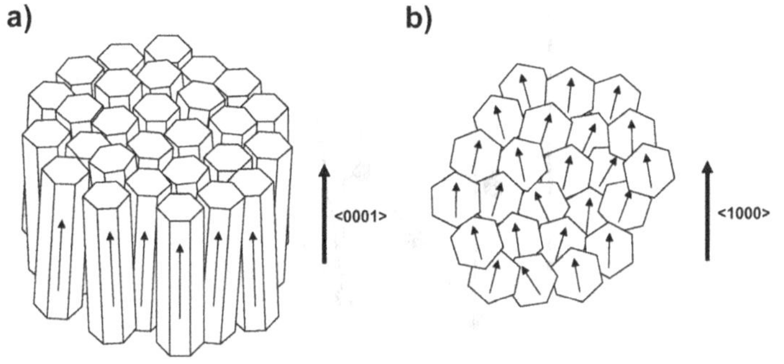

Figure 2.9. *Columnar structure of GaN: a) the disorientation of the columns relative to <0001> (tilt) generates mainly threading dislocations with mixed character (Burger vector a+c); b) the in-plane disorientation of the columns creates pure edge (Burger vector a) threading dislocations*

The columns disoriented along the [0001] axis (tilt) create threading dislocations with the screw component (Burger vector c or a+c), and the columns disoriented in the (0001) plan (twist) create threading dislocations with the edge component (Burger vector a).

Edge type dislocations (at grain boundaries) have a Burger vector resulting from a local crystal deformation with value a=1/3<1-210>, however screw dislocations have a Burger vector with a value equal to c=<0001>.

For a standard MOVPE epitaxial growth of GaN on sapphire, the typical dislocation density is ~ $10^9/cm^2$ for type a and a+c and $10^7/cm^2$ for type c.

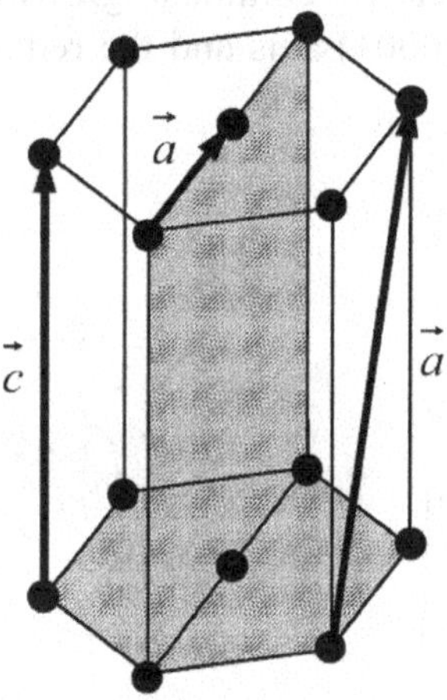

Figure 2.10. *Schematic of the three threading dislocation types in the nitride lattice: a, edge (Burger vector 1/3<11-20>); c, screw (<0001>) ; a+c, mixed (1/3<11-23>)*

X-ray rocking curves (ω-scan) from the symmetric (0002) diffraction peak are generally used to estimate the tilt of the columnar structure. The full width at half-maximum (FWHM) of the rocking curves gives some indication of the tilt.

It represents an indirect measurement of the mixed and screw dislocation density. Typical FWHM values for the (0002) reflection are between 350 and 550 arcsec when GaN is grown on sapphire.

It is also possible to directly measure the dislocation density by atomic force microscopy (AFM) on film surface.

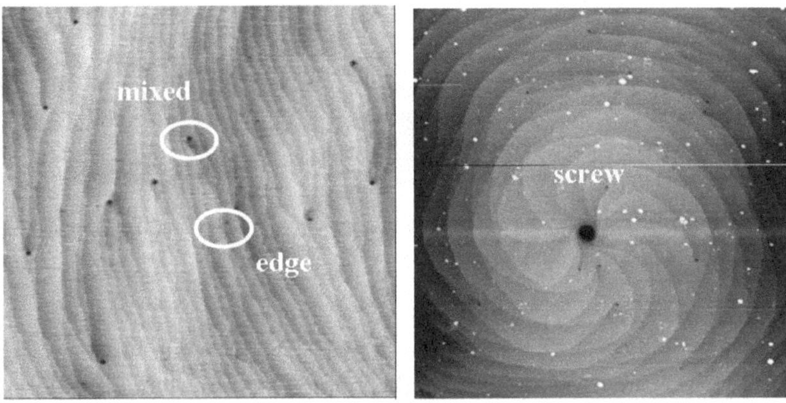

Figure 2.11. *AFM images (2x2 μm^2) of MOVPE GaN over sapphire (see color plate section)*

AFM surface analysis (Figure 2.11) shows ridges, more or less parallel, in which some mixed dislocations are rising at the intersection between two ridges. Edge dislocations, represented by small depressions, are randomly placed and harder to see. An equivalent number of mixed and edge dislocations can usually be found. Pure screw dislocations are, however, less common and noticeable by a spiral growth around the dislocation.

2.3.2. *Disorientation of the substrate*

As a general rule, the substrate surface morphology controls the growth mechanisms. In the cases where the surface plane is disoriented with respect to a normal crystal plane, the surface is formed with steps and terraces. The bigger the disorientation, the more narrow these steps and terraces tend to be. With MOVPE, high temperature growth facilitates the diffusion of atoms to the surface and their displacement from terraces to steps. When the diffusion length is longer than the terraces width a "step flow" growth mode occurs, and the resulting morphology is shown in Figure 2.11. This morphology changes when the surfaces are formed with wide terraces. Figure 2.12 shows the GaN surface evolution for different disorientation θ of the sapphire. For low values of θ (Figure 2.12a), the surface roughness is relatively high with a great number of bumps. Smoother surfaces are obtained for $\theta \sim 0.3°$ (Figure 2.12c).

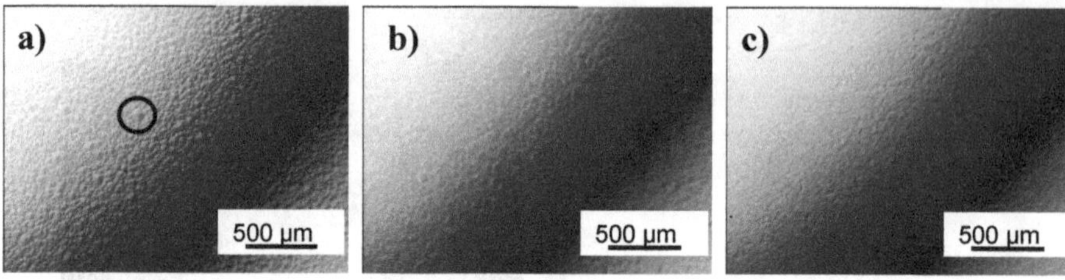

Figure 2.12. *Optical microscopy images of GaN grown on sapphire (0001) with an initial disorientation of a) $\theta = 0.15°$, b) $\theta = 0.23°$, c) $\theta = 0.31°$*

The AFM image, shown in Figure 2.13 (θ = 0.15°), shows that a screw dislocation appears at the center of the bumps with a characteristic spiral growth. The average height of the bumps is 5 nm.

These defects are observed when the nucleation starts on a terrace and the phenomenon is enhanced as the terraces become wider than the diffusion length of the surface atoms. This kinetic is generally observed with MBE growth when the surface atoms have a low diffusion speed due to the relatively low growth temperature.

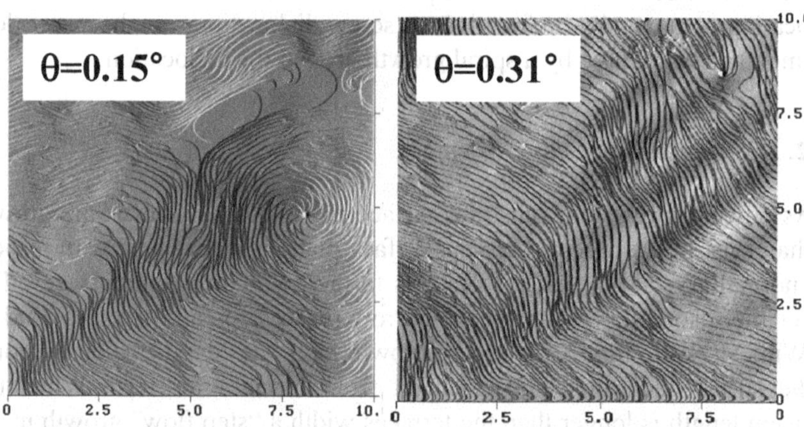

Figure 2.13. *AFM images (10x10 μm^2) of GaN surface grown on (0001) sapphire with a disorientation of $\theta = 0.15°$ and $\theta = 0.31°$. The bump (left figure) arises from a spiral growth around a screw dislocation*

2.3.3. *Epitaxial stress*

Al_2O_3 and SiC have parameter mismatches of -13.8% and -3.4% respectively with GaN; that is, if the epitaxial growth of GaN is lattice-matched with the substrate (pseudomorphic growth), compression of the GaN layer should occur causing the wafer to present a convex bow.

However, *in situ* studies show that the opposite occurs [HEA 99] and during GaN growth on sapphire, the layer is under tension, causing the wafer to present a concave bow. This deformation increases as the growing film gets thicker.

In fact, the GaN growth on sapphire starts with the coalescence of grains at the nucleation layer. Theoretical models have shown that when a film grows by coalescence, tensile stress results [FRE 01].

This stress can generate cracks for GaN thicknesses of greater than 5 µm. The constraint strongly depends on the growth condition and will be more important if there is a high density of grains in the nucleation layer.

In the same way, since the grains are twisted in the plane, their coalescence provokes some edge dislocations (see Figure 2.9). This means that there is a direct relationship between the edge dislocation density ρ_d and the tensile stress in the GaN layer.

It was experimentally shown that at growth temperature, the epitaxial tensile stress increased with ρ_d rising from 0.05 GPa at $\rho_d = 10^8/cm^2$ to 0.3 GPa at $\rho_d = 2 \times 10^9/cm^2$.

2.3.4. *Thermal stress*

During cooling after growth, important stress arises because of the differences in CTE between the substrate and GaN (see Table 2.1).

The CTE of 6H-SiC is lower than that of GaN and in this case the growing layer is under tension, creating a concave deformation. This stress can be a serious problem as it causes some cracks to form during cooling.

With Al_2O_3, the CTE is higher and the thermal stress will be compressive, causing the wafer to form a convex bow at room temperature.

Figure 2.14 shows the bow for a 4 µm GaN layer grown on a 450 µm thick Al_2O_3 substrate that is 2 inches (~50.8 mm) in diameter.

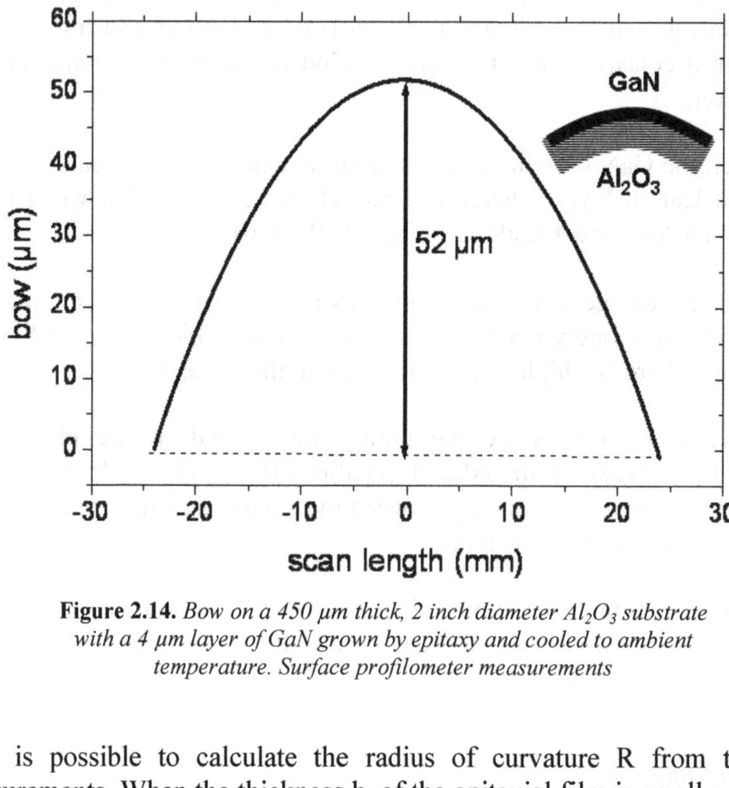

Figure 2.14. *Bow on a 450 µm thick, 2 inch diameter Al_2O_3 substrate with a 4 µm layer of GaN grown by epitaxy and cooled to ambient temperature. Surface profilometer measurements*

It is possible to calculate the radius of curvature R from the bow measurements. When the thickness h_f of the epitaxial film is small compared to the substrate thickness, the Stoney's equation [HEA 99] relates the 1/R curvature and the stress, according to:

$$\frac{1}{R} = \frac{6\sigma_f}{M_s h_s^2} h_f \qquad [2.3]$$

σ_f is the film stress and M_s is the biaxial modulus of (0001) Al_2O_3. This equation shows a linear relationship between the 1/R curvature and the film thickness h_f.

Figure 2.15 shows the curvature evolution for different thicknesses of GaN grown on a 450 μm-thick sapphire substrate.

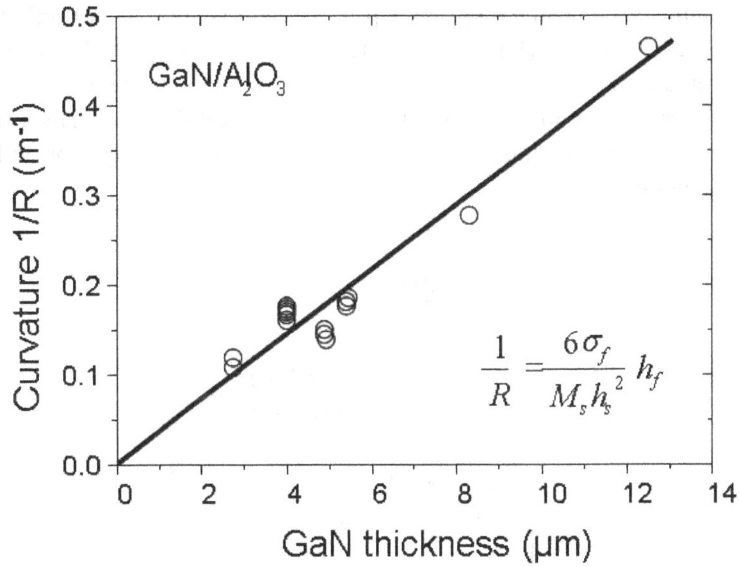

Figure 2.15. *Linear variation of the wafer curvature as a function of GaN film thickness*

Using the slope of the curve, it is possible to calculate the thermal stress. With Ms = 602 GPa (biaxial modulus of sapphire) the stress is $\sigma_f = 0.71$ GPa, in good accordance with the value of 0.71 ± 0.10 GPa [BOT 01] obtained from the determination of the GaN lattice parameters, using X-ray diffraction measurements. A compressive stress in GaN does not result in the formation of cracks; however, the convex bow of the wafer creates a tensile stress in the substrate near the interface. For a critical value, cracks may appear in the substrate and propagate through the GaN layer. This phenomenon has been observed for GaN layers over 20 μm thick [HIR 93].

2.4. MOVPE growth of GaN on sapphire

2.4.1. *GaN growth*

MOVPE was first used in the 1970s for (Al,Ga)As [MAN 72]. It is a gas phase technique which uses metalorganic M-R_n compounds, where M is the metal and R the organic radical. For GaN, the Ga source is generally trimethylgallium $Ga(CH_3)_3$ (TMG) or triethylgallium $Ga(C_2H_5)_3$ (TEG). P-doping comes from bis-cyclopentadienyl magnesium (Cp_2Mg). Most of metalorganics are liquid except for Cp_2Mg which is solid at normal ambient temperatures (0-40°C) [DEM 00]. These compounds are held at constant temperatures in hermetic bottles using a thermostatic bath. Hydrogen is used as a carrier gas which flows through the liquid and transports the metalorganic in a gas-phase into the growth chamber. NH_3 gas provides the nitrogen source.

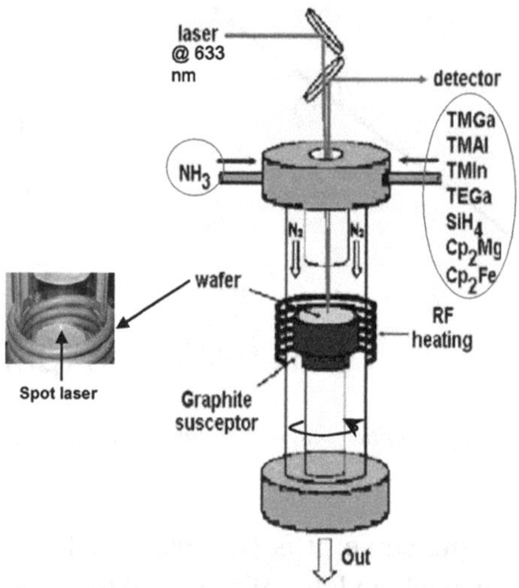

Figure 2.16. *Schematic of an MOVPE reactor. Metalorganic compounds and NH_3 are introduced into the chamber via two separate channels. The substrate is placed on a graphite susceptor heated by radio-frequency inductive coupling. The growth is controlled in situ, using a laser reflectometry setup*

Figure 2.16 is a schematic of an MOVPE reactor in vertical configuration. Metalorganic and NH_3 precursors are introduced via two separate inlets in the chamber, in order to reduce parasitic reactions in gas phase. The precursors are pyrolized on the substrate at ~1,000°C via a radio-frequency induction-heated graphite susceptor.

A constant NH_3 and metalorganic flow rate ensures the growth of nitride film on the substrate, while the hydrocarbon by-products flow through the exhaust. During the process, the pressure in the chamber is held between 10^4 and 10^5 Pa[1].

Before the formation of GaN from NH_3 and TMG, an intermediate adduct forms according to the following reaction:

$$(CH_3)_3Ga + NH_3 \rightarrow (CH_3)_3Ga : NH_3 \qquad [2.4]$$

These adducts are very unstable at the growth temperatures of 1,000-1,100°C, so that they dissociate to form $Ga_{(g)}$ and finally the overall reaction is:

$$Ga_{(g)} + NH_{3(g)} \rightarrow GaN_{(s)} + 3/2H_{2(g)} \qquad [2.5]$$

It should be noted that, thermodynamically, the dissociation of NH_3 according to:

$$NH_{3(g)} \rightarrow 1/2N_{2\,(g)} + 3/2\ H_{2\,(g)} \qquad [2.6]$$

is completed at T>300°C. However, the kinetic of reaction [2.6] is very slow compared to reactions [2.4] and [2.5], so that during MOVPE the breakdown of NH_3 into the inactive H_2 and N_2 molecules is negligible [KOU 97].

Due to the high temperatures used during growth, a large flux of NH_3 (V/III ratio>1,000) is necessary in order to avoid nitrogen evaporation from the GaN surface.

The growth is usually controlled using an *in situ* laser reflectometry [BEA 97]. The differences in the refractive index at the air/film/substrate interfaces cause multiple reflections within the film. During growth, the signal reflected by the film will experience oscillations which are either constructive (*maxima*) or destructive (*minima*). The oscillation period is given by: $\tau = \dfrac{\lambda}{2nV}$, where λ is the laser wavelength, n the refractive index of the film and V the growth velocity. The reflectivity of the signal is very sensitive to the flatness of the interfaces and the oscillation amplitude decreases as the surface roughness increases. The signal may also fully disappear when the grown material forms scattered crystals.

1. 10^5 Pa = 1 bar (atmospheric pressure).

2.4.2. *Standard 2D epitaxy*

The strong lattice mismatch between GaN and Al2O3 (-14%) produces polycrystals during growth at 1,100°C. The use of AlN [AMA 86] and GaN [NAK 91] intermediate layers was a major step in the amelioration of heteroepitaxial growth of nitrides on Al_2O_3. This nucleation layer, deposited at low temperature, creates a uniform layer on the sapphire, helping the nucleation for the subsequent growth of the GaN film deposited at high temperature. The improvement in the crystal quality as a result of the intermediate layer is seen in Figure 2.17.

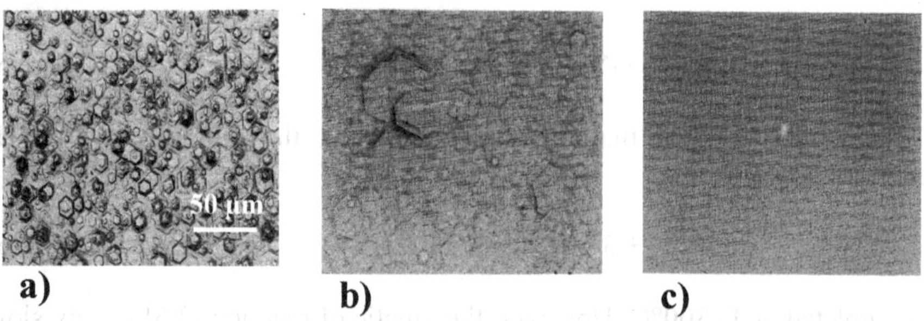

Figure 2.17. *Epitaxy of GaN at 1,030°C directly on Al_2O_3 (a), on a buffer layer of GaN deposited at 600°C for 30 s (b), and for 70 s (c). For longer deposition times the layer becomes rougher [NAK 91]*

Figure 2.18 shows the growth steps of GaN on a AlN buffer layer [HIR 91]. An amorphous layer of AlN is first deposited on Al_2O_3 at a temperature of 500 to 600°C. The temperature is then increased to 1,030°C. In the first step, AlN crystallizes into hexagonal columns with a general (0001) orientation with slight degrees of disorientation relative to the surface plane. TMG is then introduced into the reactor chamber. In the second step, GaN nucleation starts at the tip of the AlN columns. During the third step, the process becomes selective as GaN grows only on the columns with a perpendicular orientation to the surface. During growth the columns widen, forming a trapezoid (steps 4 and 5) which further coalesce to create a 2D film.

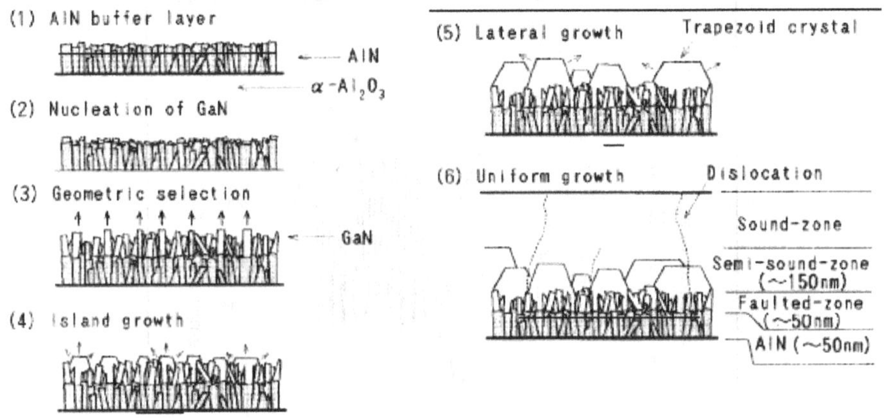

Figure 2.18. *GaN growth on a buffer layer of AlN deposited at low temperature [HIR 91]*

The same mechanisms are observed with a GaN buffer layer. The coalescence of the columns create threading dislocations. The grains with a twisted disorientation in the (0001) growth plane create edge dislocations, whereas grains with a tilted disorientation to the [0001] axial growth direction create mixed dislocations.

The total dislocation density of such epitaxial layers is on the order of 10^9-10^{10}/cm^2 [WU 98] and the linewidth (FWHM) of the X-ray rocking curves for the (0002) reflection is 350 to 550 arcsecs.

2.4.3. *3D epitaxial growth*

The defects observed in an epitaxial layer of GaN are mainly dislocations from the coalescence of different grains during the first growth step. The dislocation density is therefore directly related to the density of nucleation grains. An interesting approach [BEA 01] uses silicon which has an anti-surfactant effect in GaN [SHE 98] and reduces the density of nucleation sites in the buffer layer. Figure 2.19 shows the reflectivity signal measured *in situ* during growth.

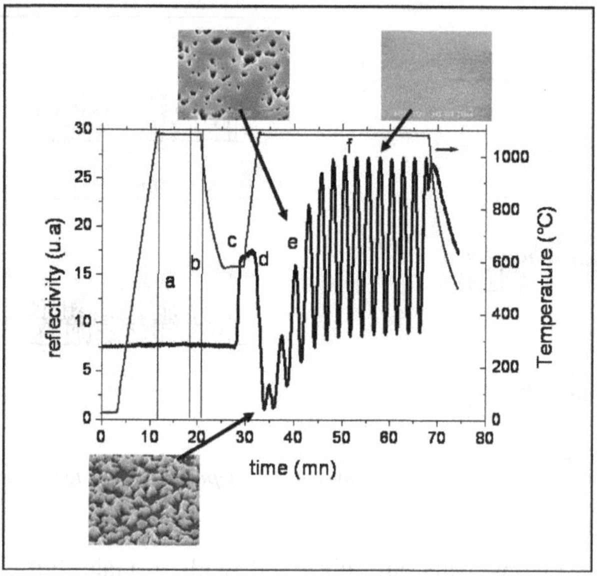

Figure 2.19. *Reflectivity signal evolution during 3D GaN growth on sapphire*

The different steps of 3D growth are as follows:

(a) a monolayer of AlN forms on the sapphire surface by the introduction of NH_3 into the chamber at 1,080°C for a few minutes;

(b) a monolayer of silicon nitride is deposited by exposing the layer to a gas flow of silane (SiH_4) and ammonia. This step is referred to as Si/N treatment;

(c) the temperature is reduced to about 600°C and a GaN buffer layer is deposited by the introduction of TMG in the reactor;

(d) the temperature is increased again to 1,080°C. During this stage, the initial 2D buffer layer experiences a major morphologic change by forming 3D GaN islands by mass transport recrystallization. This transformation occurs due to the Si/N treatment of sapphire (step b). During this 2D to 3D transformation the reflectivity signal collapses because of the increased diffused scattering of the laser spot over the surface.

TMG is then introduced into the chamber shortly after the 2D to 3D transformation. Growth occurs by lateral and vertical expansion of the islands;

(e) lateral expansion favors the coalescence of the islands. During this coalescence step, the reflected signal exhibits increasing oscillation amplitudes;

(f) the full coalescence of the grains is completed. The growth then becomes 2D and the oscillations are constant.

The resulting layer has a dislocation density of 5 x $10^8/cm^2$, which is 10 times lower than for 2D GaN. It can be further reduced by a factor of 10 [FRA 02] by increasing the duration of Si/N treatment (step b). In this case the number of GaN islands (step d) is greatly reduced and the coalescence step (step e) becomes much longer. Thickness layers of 10μm are then necessary to obtain a smooth surface without pits.

Compared to 2D epitaxy, the FWHM of (0002) reflection from X-ray diffraction is improved from 350-550 arcsecs to 180-360 arcsecs.

2.4.4. *Epitaxial lateral overgrow (ELO 1S)*

The defects created at the interface that propagate in the direction of growth constitute the main problem of heteroepitaxial growth. With ELO, the lateral growth is favored, thereby reducing these interface effects [GIB 04].

The 3D growth mode described in section 2.4.3 has some similarities with ELO since the layer is being formed by lateral coalescence of isolated GaN islands. The ELO technique is, however, more organized. The principle of this technique is illustrated in Figure 2.20:

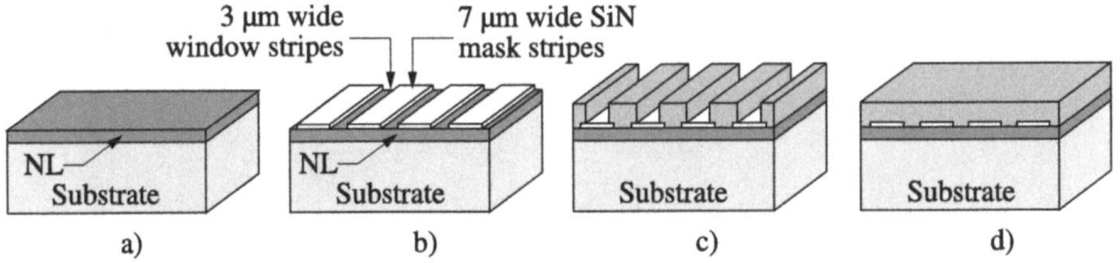

Figure 2.20. *GaN growth using ELO 1S*

(a) first, a few microns of GaN is grown epitaxially on Al_20_3;

(b) a dielectric layer of SiO_2 or Si_3N_4 is then deposited using photolithographic processes. This layer forms alternative lines of dielectric and GaN windows;

(c) during a second epitaxy, the nucleation occurs selectively at the GaN openings;

(d) GaN grows, forming stripes that further coalesce laterally above the dielectric. This technique effectively reduces the number of defects. The dislocation density above the GaN openings is identical to that of the GaN underlayer. Above the dielectric, the density is reduced; however, some defects at the coalescence boundary are observed [ROM 03].

The ELO GaN was analyzed by cathodoluminescence (CL). In this technique, the material is excited using a scanning electron beam and a mapping of the luminescence signal from the material is recorded.

On such images, the dark regions correspond to a high concentration of non-radiative defects.

Figure 2.21 shows the luminescence at λ=358 nm, that is, the emission of the free A exciton in GaN [LER 97].

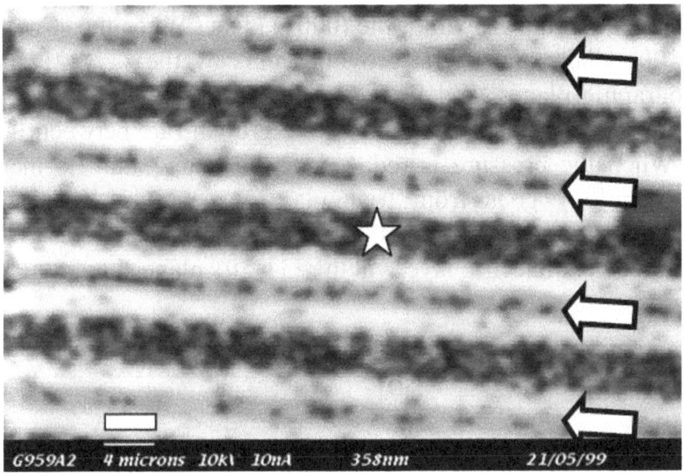

Figure 2.21. *CL image of a GaN layer using ELO 1S growth. The arrows point to the lines where the GaN coalesces above the dielectric. The dark regions, indicated by the star, correspond to the growth of GaN above the window. The width of the white bar = 4 μm [BEA 99]*

The dark spots correspond to the emergence point of dislocations. The spatial resolution is limited by the size of the exciting electron beam. The bright stripes correspond to the coalescence of GaN above the dielectric. The dark stripes represent the nucleation of the GaN layer above the initial GaN underlayer. In these dark regions the dislocation density is approximately $5x10^8/cm^2$. In the bright zones the density is reduced to $5 \times 10^6/cm^2$. This image is a good example of the heterogenous distribution of the dislocations on the surface obtained by ELO 1S and the efficient reduction of the dislocation density.

2.4.5. *Anisotropic growth*

Grain coalescence occurs when the lateral growth dominates the vertical growth. ELO growth depends on the orientation of the stripes with respect to the crystallographic direction of the GaN underlayer [HIR 02]. Figure 2.22 illustrates the morphologic differences in the resulting GaN layer grown by ELO, depending on the mask orientation. When the stripes are parallel to <11-20>, the transversal cross-section of GaN, grown using ELO, shows triangular aspects regardless of the growth conditions (Figure 2.22a). The coalescence of the grains is difficult since the growth of the {1-101} plane is slow compared to the growth of the (0001) plane. When the stripes are along the <1-100> plane, the lateral growth occurs by the {11-22} and {11-20} crystallographic planes, and the vertical growth by the {0001} plane. The GaN morphology depends on the relative growth rate between the different planes. When the pressure is reduced or the temperature is increased, the GaN morphology goes from triangular to rectangle. The sides of the rectangles are defined by the slow crystallographic planes. Coalescence occurs above the dielectric layer via the {11-20} planes junction.

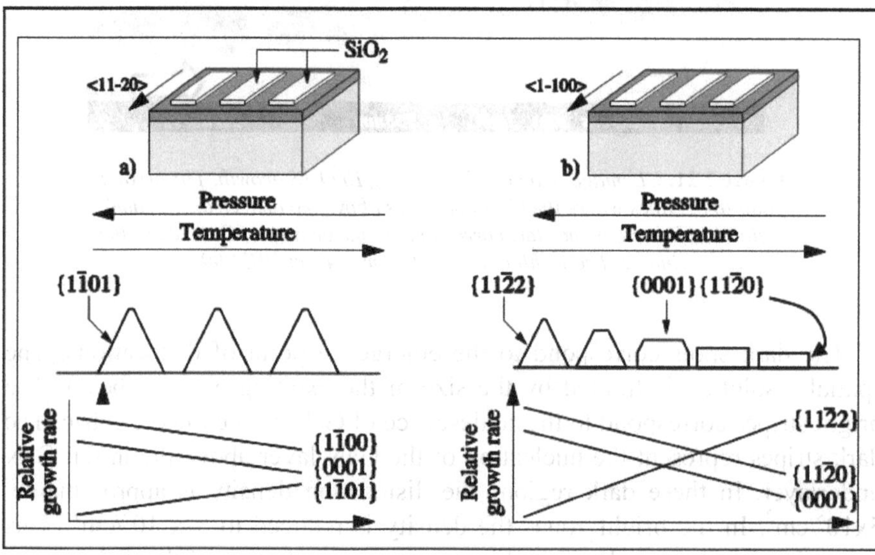

Figure 2.22. *Schematic evolution of ELO GaN grown on <11-20> (a), and <1-100> (b) oriented stripes. The temperature and pressure influence on the GaN morphology is also shown [HIR 02]*

It is therefore possible to increase the lateral growth while decreasing the vertical growth of the ELO stripes by increasing the temperature or decreasing the pressure. The same phenomenon may also be achieved by introducing Mg vapor during growth. In this case it is possible to create ELO layers without modifying the pressure or temperature [BEA 98].

The ability to modify the geometry of the GaN stripes by adjusting the growth conditions can be used to further improve the quality of the films.

This is the case with ELO 2S, which is a technique based on a two stage GaN growth [BEA 99].

2.4.6. *Two stage ELO GaN growth (ELO 2S)*

The two stage ELO technique starts with growth conditions that facilitate the vertical growth compared to the lateral growth (low temperature at atmospheric pressure), which leads to grains which are perfectly triangular in cross-section (Figure 2.23b). The second stage uses high temperature and low pressure conditions to develop lateral growth until full coalescence of the grains is achieved (Figures 2.23c and d)

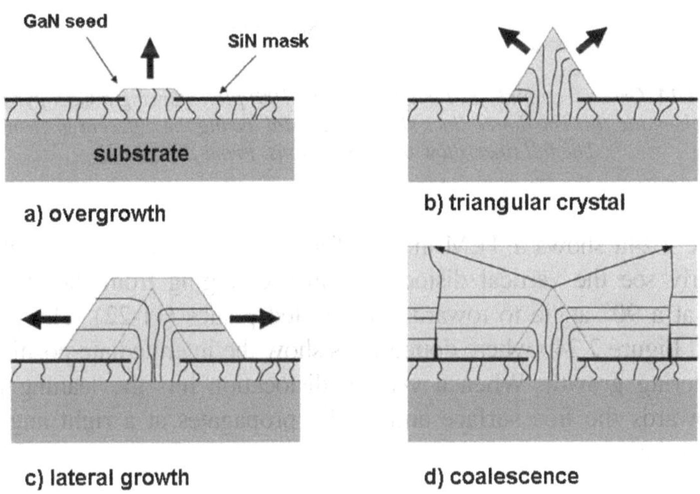

Figure 2.23. *Cross-section of an ELO 2S at different growth stages. At re-growth the grains have a trapezoidal shape a), at first they develop with a triangular shape with some curved dislocations b), then the growth becomes lateral c), until coalescence of the grains d). Drawings from [FEL 02]*

When the coalescence is complete, the only remaining dislocations are those at the coalescence boundaries as well as those dislocations propagating vertically throughout the entire process, that is, the dislocations situated at the tip of the triangles which emerged at the surface. This ELO technique very efficiently reduces the dislocation density by several orders of magnitude, from $5x10^8/cm^2$ in the starting layer to $10^6/cm^2$. However, a 7 µm grown thickness is necessary to obtain full coalescence on sapphire.

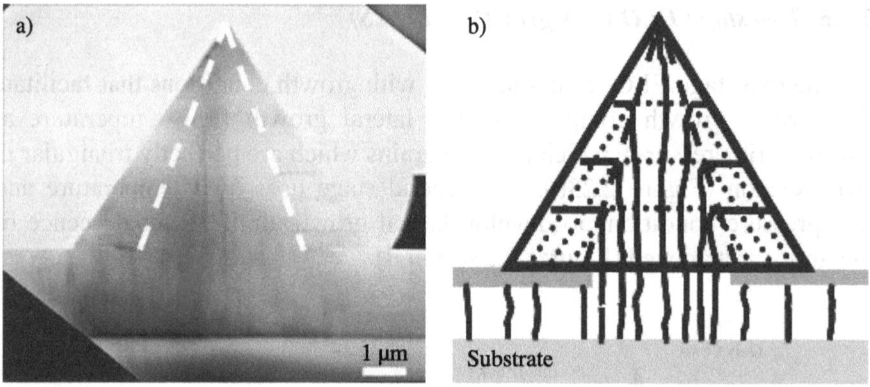

Figure 2.24. *Cross-sectional TEM image of the (1-100) plan showing a GaN pyramid (a) and its schematic (b). The dotted lines show the crystal during the different growing stages. The full lines show the dislocations. From [VEN 00]*

Figure 2.24a shows a TEM image of a GaN pyramid in cross-section. We can clearly see the vertical dislocation lines emerging from the surface and bending at a 90° angle to towards the faceted planes (11-22). The process is described Figure 2.24b where dotted lines show the intermediate position of the crystal during growth. When a vertical dislocation hits the leaning plane, it bends towards the free surface and finally propagates at a right angle in the (0001) basal plane.

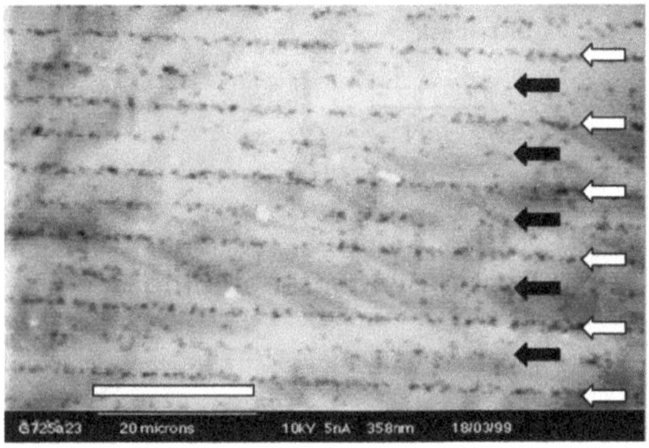

Figure 2.25. *CL image (λ=358 nm) of the GaN surface grown by ELO 2S (to be compared with ELO 1S growth in Figure 2.21). The white arrows define the coalescence boundary above the dielectric. The black arrows define the open area where the dislocations propagate vertically from the substrate. Length of the white bar = 20 μm [BEA 99]*

The cathodoluminescence image (Figure 2.25) of an ELO 2S GaN surface shows parallel dislocation lines which are caused by the geometry of the dielectric mask. The coalescence zones (white arrows) display a high dislocation density. At the middle of the opened zones, the dislocations issued from the first growth step are situated at the pyramid apex and propagate to the surface. The average dislocation density of an ELO 2S layer is about $5 \times 10^7/cm^2$. The width of the white areas is 10 µm which is sufficient for optoelectronic devices. Using substrates obtained by a combination of ELO re-growth and HVPE (hybrid vapor phase epitaxy) thickening, the company Nichia achieved laser diodes with a 15,000 hour lifetime and 30 mW power [NAG 00].

2.4.7. *GaN growth using pendeo-epitaxy*

Simultaneously with classical ELO growth, another technique which only uses lateral coalescence was developed: pendeo-epitaxy [DAV 01]. The goal of this technique is to optimize the zone where the dislocation density is low and the GaN is of high quality. It is achieved by starting the growth exclusively on vertical walls.

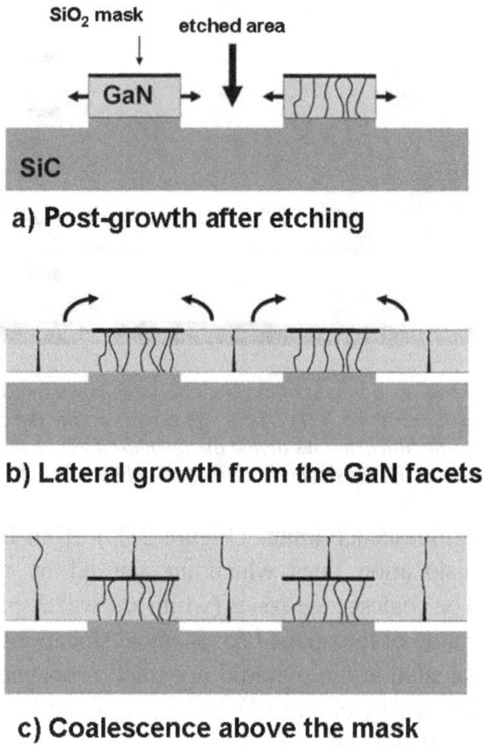

Figure 2.26. *The pendeo-epitaxial growth of GaN on SiC, from [DAV 01]. After etching the first GaN layer a), the germs grow laterally above the SiO_2 mask b) until full coalescence c)*

The principle of pendeo-epitaxy is shown Figure 2.26. First, a SiO_2 mask is deposited over a layer of GaN, then stripes are etched through the substrate, which also prevent the GaN nucleation at high temperature. The grains can only grow laterally until they meet, then a classical ELO growth takes place. The lateral layer expands above the SiO_2 surface which had been deposited on the GaN layer. In this way all the elements from the second growth only develop laterally, which in theory should eliminate all the dislocations in the final layer except for those at the coalescence boundaries. As with ELO 2S, the dislocations are located at the coalescence boundaries, in parallel lines with a pitch defined by the mask. A cross-section of such a layer is shown in Figure 2.27.

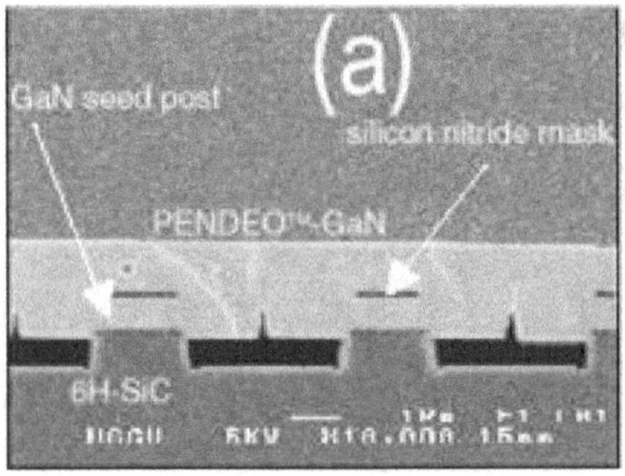

Figure 2.27. *SEM image of GaN layer cross-section, grown using pendeo-epitaxy, from [DAV 01]. Some voids can be seen under the coalescence zone*

Variations on this technique were also used, such as pendeo-epitaxy without SiO_2 masks [GEH 00], or selective pendeo-epitaxy, referred to as "air-bridge" ELO [KID 00].

Although the technologies used with pendeo-epitaxy are more complex than with ELO, it is more effective for reducing the dislocation density (between 10^6 and 10^7 dislocations/cm^2) even with thinner layers, because in this case the ratio between lateral growth and vertical growth is four times higher.

2.4.8. *Nano epitaxy*

When there is a high parameter mismatch in 2D epitaxy, a critical thickness exists, where the grown film starts to release the elastic energy absorbed which creates some dislocations. With GaN/Al_2O_3, this phenomenon occurs at the first monolayer. When the epitaxial growth occurs over nanometric surfaces, on the order of this critical thickness, it is theoretically possible that no dislocations may develop in the layer [LUR 86]. This concept was verified by growing nanocolumns of GaN without defects, using nanolithography techniques [HER 06], by a catalyst assisted nucleation of GaN over metallic dots [WAN 06] or directly on silicon [CAL 00].

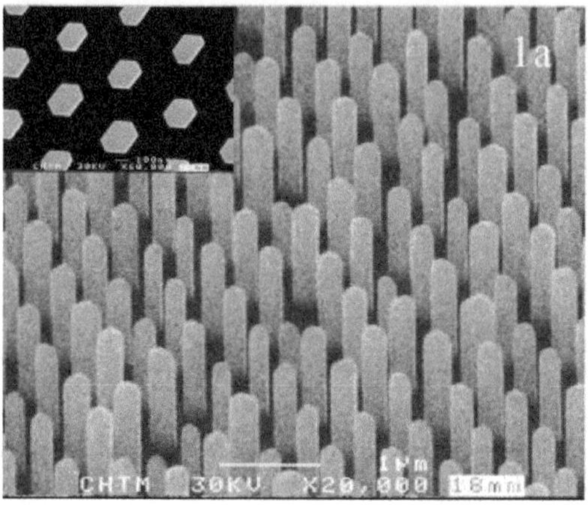

Figure 2.28. *Nanocolumn array of 1 μm-high GaN. MOVPE growth from a nanolithography mask. The inserted image is a top-down view of the columns which exhibit a hexagonal structure [HER 06]*

Recently, these nano-columns were used to achieve nitride nano-LEDs [KIK 04, KIM 04]. An ELO growth of GaN using AlN/6H-SiC nano-pillars was also achieved (Figure 2.29) in order to grow a GaN coalesced layer with a low dislocation density [BOU 06]. The morphologic evolution of the coalescence should lead to a a flat film with a dislocation density of $< 10^6/cm^2$.

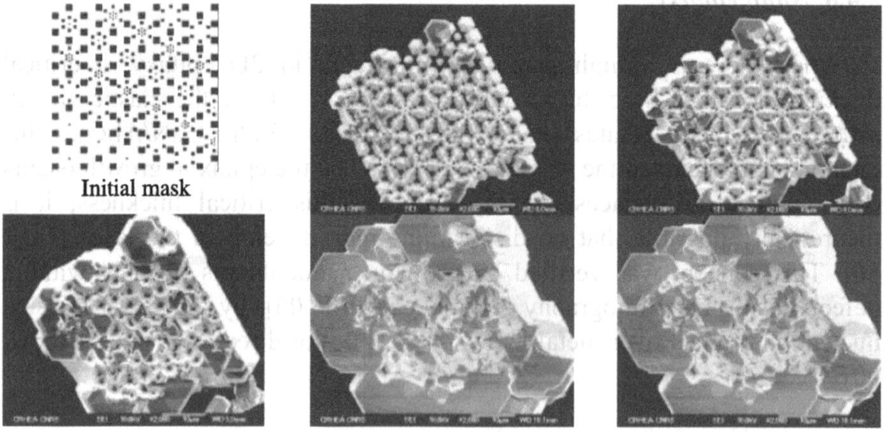

Figure 2.29. *MOVPE re-growth over nano-pillars of AlN/SiC. The lithographic mask (160x160 μm^2) is designed with square motifs, the smallest of which are 40 nm per side [BOU 06]*

2.5. Bulk nitride substrates

Nitrides sublimate at much lower temperatures than their melting temperature, therefore it is not possible to use standard methods such as Czochralski or Bridgmann, which use a stoichiometric bath at atmospheric pressure, to synthesize them.

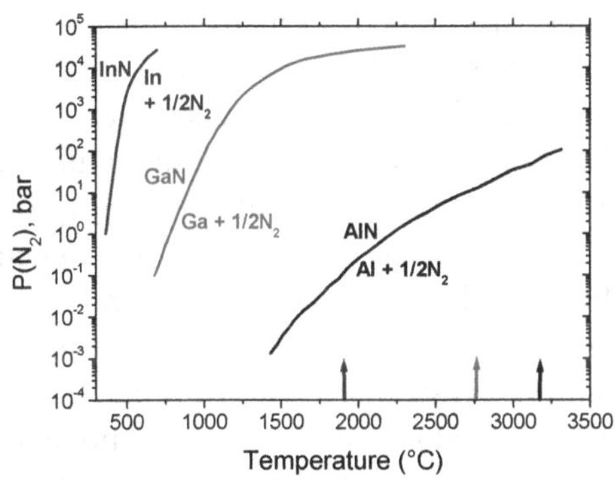

Figure 2.30. *N_2 pressure in equilibrium with InN, GaN, and AlN, adapted from [GRZ 95]. The vertical arrows give the melting temperatures, T_M = 1,873°C (InN) , 2,518°C (GaN) and 3,214°C (AlN), calculated from the III-N bond energy*

Figure 2.30 shows the equilibrium nitrogen pressure over nitride surfaces as a function of temperature. Depending on the compounds, strong variations are observed in the stability of nitrides. For a 1 bar ambient pressure of N_2, AlN remains stable up to 2,250°C, GaN up to 800°C and InN starts degrading at around 400°C. It can also be seen that InN may reach its melting temperature at a liquid state only under a nitrogen pressure over 100 kbar, which cannot be achieved experimentally. For AlN, a 70 kbar nitrogen pressure is necessary to reach an equilibrium point at the melting temperature, which may be achieved experimentally; however, the melting temperature in this case becomes the limiting factor. GaN with a 45 kbar N_2 pressure at equilibrium at the melting temperature is in a intermediate situation.

2.5.1. *HNPS (high nitrogen pressure solution method) for the fabrication of crystalline GaN*

The first experiments to grow GaN from liquid gallium and gaseous ammonia were performed in 1930. A GaN layer is created after a few hours at 900-1,000°C. However, the resulting GaN quality is not very good since at these temperatures Ga evaporates and GaN sublimates. High pressure growths were later developed. Finally, under conditions near the thermodynamic equilibrium (high pressure and high temperature), some bulk GaN monocrystals were obtained with centimeter scale dimensions [GRZ 01]. A liquid gallium bath is heated to a temperature of 1,700°C under 20 kbar nitrogen. Under these conditions, the solubility of atomic nitrogen in the gallium bath reaches 5 at. %. The crystallization occurs spontaneously in the cold parts of the crucible where the temperature gradient is under control. The crystals are then separated from the gallium using a boiling nitric acid treatment. The resulting crystals are 100 μm thick and oriented towards the (0001) axis (see Figure 2.31) with a smooth N polar face and a rough Ga face. The dislocation density is very low at ~$2 \times 10^2/cm^2$, and the FWHMs of the (0004) X-ray rocking curves are as low as 40-50 arcsecs, compared to 250 arcsecs obtained with the epitaxial growth of GaN over sapphire. The resulting material is highly n-type doped with a 60 cm^2/Vs mobility, due to the oxygen contamination in the chamber. Resistive GaN may be obtained by doping compensation with Mg or Be [KRU 01].

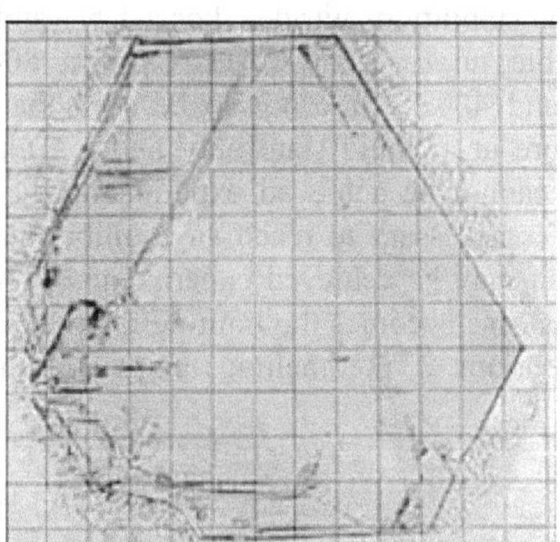

Figure 2.31. *GaN crystal grown by HNPS (1 mm grid) [GRZ 01]*

A variation of this method consists of adding sodium (Na) to a liquid Ga bath [AOK 00]. The drastic conditions on the temperature and the pressure are reduced thanks to the catalysis of Na on the N_2 dissociation at the Ga bath interface. GaN growth is then possible at 750°C and 50 bars. The final crystals are colorless and transparent, approximately 1 mm^2, n-type doped at 1-2 x $10^{18}/cm^2$ and with a mobility of 100 $cm^2/V.s$. The FWHM of the (0004) X-ray rocking curve is 25 arcsecs.

The small crystal size and the slow growth (< 1 µm/h) prevent the use of this method for industrial purposes. To increase the crystal size, sapphire and SiC substrates were used to grow GaN by HNPS [BOC 04]. Polycrystalline GaN is obtained with sapphire, probably because of the strong parameter mismatch (13%) between the two materials. Growth over SiC leads to cracks in the material during cooling.

The best results are obtained when GaN is first deposited over sapphire using MOVPE. The surface is made of macro-steps and the dislocation density is about 5 x $10^7/cm^2$. The displacement of the crystal front, which defines the growth speed, depends on the axial temperature gradient. Maximum velocities of 2 µm/h are required to avoid instabilities such as polarity inversion or to limit roughness.

2.5.2. *Ammonothermal synthesis of GaN*

This synthesis follows the hydrothermal method, which is used with quartz and has been successfully applied to ZnO as well [OSH 04]. Compared to HNPS, where elemental nitrogen is dissolved in Ga, the ammonothermal method dissolves Ga or GaN in a supercritical NH_3 solution. GaN is not directly soluble in NH_3, and mineralizers such as sodium amide ($NaNH_2$) [HAS 05] or NH_4Cl [KAG 06] are necessary to dissociate NH_3 and to form a soluble gallium amide $Ga(NH_2)n$. The growth is achieved in an autoclave, full of liquid ammonia at high pressure, typically between 1 to 3 kbars and at a temperature between 500 and 650°C. One end contains the dissolving compound (Ga or polycrystal GaN) and the other end the monocrystal GaN seed. A temperature gradient along the autoclave achieves the dissolution, the amide transportation and the GaN crystallization on the seed.

Recently, some 1 inch GaN monocrystals were achieved using ammonothermal method on the front side and the back side of a GaN HVPE substrate [HAS 05].

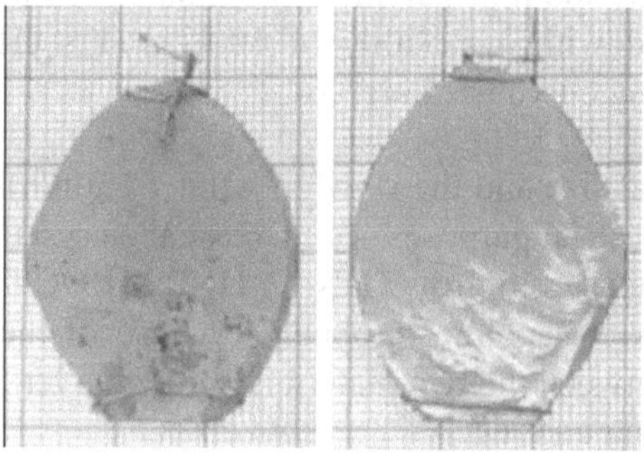

Figure 2.32. *Ammonothermal deposition of GaN on both sides of a HVPE substrate. Ga side (a) and N side (b). Grid of 1 cm [HAS 05]*

The Ga side is rough with 1 to 10 µm diameter holes; the N side is smooth with an average roughness of 3.5 nm, as measured by AFM. The dislocation density of the N layer is $\sim 10^9/cm^2$, although it is only $\sim 10^7/cm^2$ in the HVPE starting substrate. The FWHM of the (002) rocking curve from the Ga side is 2000 arcsecs and 1,700 arcsecs from the N side. These high values prove the poor quality of the layer.

The ammonothermal synthesis is still in the developing stages and some improvements are expected in the near future. The growth speeds are around 2 µm/h, which is compatible with industrial needs.

2.5.3. *Halide vapor phase epitaxy (HVPE) of GaN*

This technique, developed in the 1960s, is today the best for large surface growth. It uses an HCl gas which flows above a cup filled with liquid metallic gallium. The GaCl halide is then formed. This gas is transported to the wafer where it mixes with NH_3 to obtain GaN according to the following chemical reaction:

$$GaCl_{(g)} + NH_{3(g)} \rightarrow GaN_{(s)} + HCl_{(g)} + H_{2\,(g)} \qquad [2.7]$$

An external oven heats both the growth chamber and the boat containing the liquid gallium. The temperature and pressure conditions are the same as with an MOVPE growth, T ~ 500-1,100°C and P ~ 10^4-10^5 Pa. 100 μm/h growth rates may be achieved with this method.

Direct growth on sapphire substrates is possible, but the first nucleation steps are best controlled with MOVPE. Hence, 2 inch GaN wafers grown with MOVPE are used as the starting material.

As the layer thickens, there is a great reduction in the dislocation density. Indeed, during growth some threading dislocations bend and meet in order to cancel each other out and create loops. Even if the dislocation lines are usually parallel to the growth direction, at high speed growth the creation of faceted plans in the growing front occurs and deflects these dislocation lines, as in the ELO 2S technique.

Figure 2.33 illustrates the reduction of the dislocation density as the GaN layer deposited by HVPE on sapphire thickens. The efficiency of this method is reduced as the layer thickens since the distance between the dislocations increases. Layers of 300 μm are necessary to obtain a dislocation density of ~10^7/cm^2.

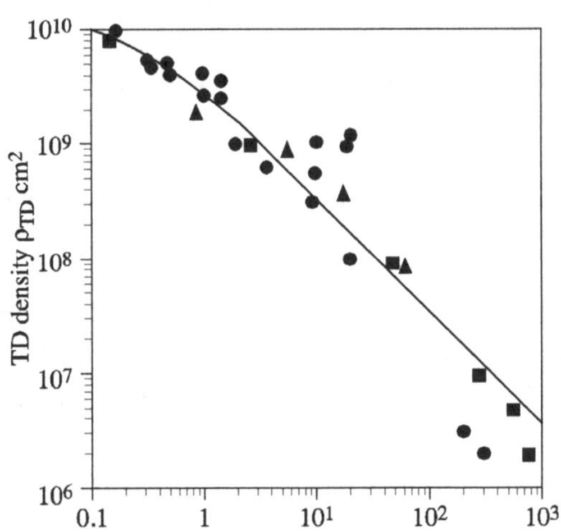

Figure 2.33. *Threading dislocation densities in GaN on sapphire grown by HVPE as a function of thickness, from [MAT 01]*

The undisputed advantage of HVPE is the ability to grow thick layers on 2 inch substrates while reducing the dislocation density. However, all the stress problems due to epitaxial growth and induced during cooling are enhanced as the thickness is increased. The tensile stress of GaN on sapphire can be greatly reduced by using a metallic gallium film deposited directly on the substrate [DAM 06]. With this technique, 100 µm thick layers grown on 2 inch substrates were achieved without cracks. However, in this case the substrates present a convex bow due to the thermal stresses during cooling.

For the formation of bulk GaN substrates, it is necessary to split the GaN layer from its substrate. Because sapphire does not dissolve easily with chemicals, some splitting processes have been developed over the past few years. A laser lift-off process uses the irradiation of the interface layer by an excimer KrF (258 nm) laser through the sapphire substrate [KEL 99]. Sapphire is transparent to this wavelength while the interface layer absorbs the radiation and thermally dissolves in the following way:

$$GaN_{(s)} \rightarrow Ga_{(g)} + 1/2N_{2\,(g)} \qquad [2.8]$$

The scanning of the laser beam over the 2 inch wafer ensures the complete separation of the GaN layer from the substrate. This method is used by Samsung-Corning in Korea, which sells such substrates. The dislocation density specified is very low ($\sim 10^6/cm^2$).

Sumitomo in Japan also markets bulk GaN with dislocation densities $< 10^6/cm^2$ [MOT 02] (http://www.sei.co.jp). These substrates are achieved using ELO HVPE epitaxy on (111) GaAs substrates. GaAs has the advantage of having a better CTE matching with GaN than sapphire does. The GaAs substrate is then removed using chemical dissolution.

In both cases, the GaN and its substrate are split after both the growth and cooling steps. These stages are critical since the mechanical stresses in the wafers are very high during growth and cooling, causing a number of wafers to break.

With the laser lift-off technique, the thermal dissolution releases some nitrogen, which creates some local stresses and can break the layer before it is fully separated from its substrate.

The heteroepitaxy research center (CRHEA-CNRS, France) [FEL 04] has recently developed a technique that uses a sacrificial layer which progressively and uniformly sublimates when a specific grown thickness is achieved. Separation occurs in the epitaxial chamber during growth. CRHEA optimized the process to deposit this sacrificial layer on sapphire or SiC and to regrow GaN over this layer using MOVPE. This technique was then transferred to LUMILOG for the growth of thick layers of GaN using HVPE. LUMILOG (http://www.lumilog.com) manufactures bulk 2 inch GaN substrates with a dislocation density of $5x10^6/cm^2$ using this process.

2.6. Conclusion

The first nitride electroluminescent diodes appeared in 1990 and their performances were surprisingly good considering the number of defects in the material. This phenomenom was later explained by indium segregation within the active layer of (In,Ga)N. However, with high power applications (lighting, lasers), these defects become an issue. LEDs are deposited by epitaxial growth on some microns-thick layers of GaN, which are grown by heteroepitaxy on Al_2O_3 or SiC substrates. LED performance is directly related to the quality of the initial GaN layer.

With heteroepitaxy, the major defects are threading dislocations stemming from the initial growth steps, which propagate through to the surface in the growth direction. It is therefore necessary to reduce this type of defect. With ELO, these defects were reduced by a factor of 100, from a dislocation density of $5x10^8/cm^2$ to $5x10^6/cm^2$ for ELO 2S GaN. These processes are complicated techniques, using a lithographic step and two growing steps, but these are the requirements necessary to achieve (In,Ga)N lasers with a 10,000 hour lifetime. Ideally, the epitaxial growth of optoelectronic devices should be achieved on bulk nitride substrates to eliminate all defects due to parameter mismatches, thermal stresses, epitaxy, etc. Nitrides cannot be grown using the Czochralski method (stoichiometric liquid bath ingots) because the nitrogen pressures at equilibrium are too high. Some techniques, using high N_2 pressure, achieved very high GaN crystal quality; however, the layer dimensions were too small (a few millimeters) for any industrial uses. Larger GaN monocrystals (~1 inch) were achieved with ammonothermal synthesis, which uses liquid NH_3. However, too many defects are still present in the final crystals. This technique is not yet mature but is the subject of much attention. The present research could lead to bulk GaN and AlN in the near future, in the same way as SiO_2 and ZnO were achieved with hydrothermal synthesis.

The vapor phase HVPE growing technique, where thick layers are grown using epitaxy on Al_2O_3 or GaAs, remains the only method to achieve large bulk substrates (50 mm). The layers are then separated from their substrate by chemical dissolution (with GaAs) [SUMITOMO process] or by laser dissolution (with sapphire) [SAMSUNG process]. Another method uses a sacrificial layer between sapphire and GaN which then sublimates during growth in the HVPE chamber and leads to a free-standing GaN layer [LUMILOG process]. These three companies are today able to produce and commercialize their products. The cost of a 2 inch GaN substrate, 400 µm thick, is about €1,500. Moving towards a larger scale industrial production will reduce the cost and achieve a turning point in the performance of nitride LEDs for lighting.

2.7. Bibliography

[AMA 86] AMANO H., SAWAKI N., AKASAKI I., TOYODA T., "Metalorganic vapor phase epitaxial growth of a high quality GaN film using an AlN buffer layer", *Applied Physics Letters,* no.48, p. 353-355, 1986.

[AOK 00] AOKI M., YAMANE H., SHIMADA M., SEKIGUCHI T., HANADA T., YAO T., SARAYAMA S., DISALVO F.J., "Growth of GaN single crystals from a Na-Ga melt at 750° C and 5 Mpa of N_2", *Journal of Crystal Growth,* no.218, p. 7-12, 2000.

[BEA 97] BEAUMONT B., VAILLÉ M., BOUFADEN T., EL JANI B., GIBART P., "High quality GaN growth by MOVPE", *Journal of Crystal Growth,* no.170, p. 316-320, 1997.

[BEA 98] BEAUMONT B., HAFFOUZ S., GIBART P., "Magnesium induced changes in the selective growth of GaN by metalorganic vapour phase epitaxy", *Applied Physics Letters,* no. 72, 921, 1998.

[BEA 99] BEAUMONT B., BOUSQUET V., VENNEGUÈS P., VAILLÉ M., BOUILLÉ A., GIBART P., DASSONNEVILLE S., AMOKRANE A., SIEBER B., "A two-step method for epitaxial lateral overgrowth of GaN", *Physica Status Solidi,* A 176, p. 567 1999.

[BOC 04] BOCKOWSKI M., GRZEGORY I., KRUKOWSKI S., LUCZNIK B., WROBLEWSKI M., KAMLER G., BORYSIUK J, KWIATKOWSKI P., JASIK K., POROWSKI S., "Deposition of bulk GaN from solution in gallium under high N_2 pressure on silicon carbide and sapphire substrates", *Journal of Crystal Growth,* no. 270, p. 409-419, 2004.

[BOT 01] BOTTCHER T., EINFELD S., FIGGE S., CHIERCHIA R., HEINKE H., HOMMEL D., SPECK J.S., "The role of high-temperature island coalescence in the development of stresses in GaN films", *Applied Physics Letters,* no.78, 1976, 2001.

[BOU 06] BOUGRIOUA Z., GIBART P., Study at CRHEA, in collaboration with the University of Cambridge.

[CAL 00] CALLEJA E., SANCHEZ-GARCIA M. A., SANCHEZ F. J., CALLÉ F., NARANJO F.B., MUNOZ E., JAHN U., PLOOG K., "Luminescence properties and defects in GaN nanocolumns grown by molecular beam epitaxy", *Physical Review B,* no.62, 16827-16834, 2000.

[CHO 03] CHOWDHURY A., NG H.M., BHARDWAJ M., WEIMANN N.G., "Second-harmonic generation in periodically poled GaN", *Applied Physics Letters*, no. 83, p. 1077-1079, 2003.

[DAM 06] DAM C.E.C., GRZEGORCZYK A.P., HAGEMAN P.R., LARSEN P.K., "Method for HVPE growth of thick crack-free GaN layers", *Journal of Crystal Growth,* no.290, p. 473-478, 2006.

[DAV 01] DAVIS R.F., GEHRKE T., LINTHICUM K.J., RAJAGOPAL P., ROSKOWSKI A.M., ZHELEVA T., PREBLE E.A., ZORMAN C.A., MEHREGANY M., SCHWARZ U., SCHUCK J., GROBER R., "Review of pendeo-epitaxial growth and characterization of thin films of GaN and AlGaN alloys on 6H-SiC(0001) and Si(111) substrates", *MRS Internet Journal of Nitride Semiconductor Research*, vol. 6, no.14, 2001.

[DEM 00] DE MIERRY P., BEAUMONT B., FELTIN E., SCHENK D., GIBART P., JOMARD F., RUSHWORTH S., SMITH L., ODEDRA R., "Influence of the Mg precursor on the incorporation of Mg in MOVPE grown GaN", *MRS Internet Journal of Nitride Semiconductor Research*, vol. 5, no.8, 2000.

[FEL 01] FELTIN S., DALMASSO P., DE MIERRY B., BEAUMONT H., LAHRÈCHE A., BOUILLÉ GIBART P., "Green light-emitting diodes grown on Si(111) by metalorganic vapor phase epitaxy", *Japanese Journal of Applied Physics*, no.40, L738, 2001.

[FEL 02] FELTIN E., "Hétéro-épitaxie de nitrure de gallium sur silicium (111) et applications", Thesis, University of Nice-Sophia Antipolis, January 2003.

[FEL 04] FELTIN E., BOUGRIOUA Z., NATAF G., Patent no. FR0311296, PCT/FR2004/02416, 2004.

[FIT 91] FITZGERALD E.A., *Dislocations in Strained Layer Epitaxy: Theory, Experiment and Applications*, North Holland, Amsterdam, 1991.

[FRA 02] FRAYSSINET E., BEAUMONT B., FAURIE J.P., GIBART P., MAKKAI ZS., PECZ B., LEFEBVRE P., VALVIN P., "Micro epitaxial lateral overgrowth of GaN/sapphire by metal organic vapour phase epitaxy", *MRS Internet Journal of Nitride Semiconductor Research,* vol. 7, no.8, 2002.

[FRE 01] FREUND L.B., CHASON E., "Model of stress generated upon contact of neighboring islands on the surface of a substrate", *Journal of Applied Physics,* no. 89, p. 4866, 2001.

[GEH 00] GEHRKE T., LINTHICUM K.J., RAJAGOPAL P., PREBLE E.A., DAVIS R.F., "Advanced pendeoepitaxy of GaN and $Al_xGa_{1-x}N$ thin films on SiC(0001) and Si(111) substrates via metalorganic chemical vapor deposition", *MRS Internet Journal of Nitride Semiconductor Research 5S1*, W2.4, 2000.

[GIB 04] GIBART P., "Metal organic vapor phase epitaxy of GaN and lateral overgrowth", *Reports on Progress in Physics*, vol. 67, no. 5, p. 667-715, 2004.

[GRA 98] GRANDJEAN N., MASSIES J., VENNÉGUÈS P., LEROUX M., DEMANGEOT F., RENUCCI M., FRANDON J., "Molecular-beam epitaxy of gallium nitride on (0001) sapphire substrates using ammonia", *Journal of Applied Physics,* no. 83, p. 1379-1383, 1998.

[GRZ 95] GRZEGORY I., JUN J., BOCKOWSKI M., KRUKOWSKI ST., WROBLEWSKI M., LUCZNIK B., POROWSKI S., "III-V nitrides thermodynamics and crystal growth at high N_2 pressure", *Journal of Physics and Chemistry Of Solids*, no. 56, p. 639-647, 1995.

[GRZ 01] GRZEGORY I., "High nitrogen pressure growth of GaN crystals and their applications for epitaxy of GaN-based structures", *Material Science and Engineering B*, no. 82, p. 30-34, 2001.

[GUE 07] GUEHNE T., BOUGRIOUA Z., VENNÉGUÈS P., LEROUX M., ALBRECHT M., "Cathodoluminescence spectroscopy of epitaxial lateral-overgrown non polar (11-20) and semipolar (11-22) GaN in relation to microstructural characterization", *Journal of Applied Physics,* no.101, p. 113101, 2007.

[HAS 05] HASHIMOTO T., FUJITO K., SAITO M., SPECK J.S., NAKAMURA S., "Ammonothermal growth of GaN on an over-1-inch seed crystal", *Japanese Journal of Applied Physics,* no.44, L1570-L1572, 2005.

[HEA 99] HEARNE E.C., CHASON E., HAN J., FLORO J.A., FIEGIEL J., HUNTER J., AMANO H., TSONG I.S.T., "Stress evolution during metalorganic chemical vapour deposition of GaN", *Applied Physics Letters*, no. 74, p. 356-358, 1999.

[HEL 98] HELLMAN E.S., "The polarity of GaN: a critical review", *MRS Internet Journal of Nitride Semiconductor Research*, no.3, 11, 1998.

[HER 06] HERSEE S. D., SUN X., WANG X., "The controlled growth of GaN nanowires", *Nano Letters*, no. 6, p. 1808-1811, 2006.

[HIR 91] HIRAMATSU K., ITOH S., AMANO H., AKASAKI I., KUWANO N., SHIRAISHI T., OKI K., "Growth mechanism of GaN grown on sapphire with AlN buffer layer by MOVPE", *Journal of Crystal Growth,* no.115, p. 628-633, 1991.

[HIR 93] HIRAMATSU K., DETCHPROHM T., AKASAKI I., "Relaxation mechanism of thermal stresses in the heterostructure of GaN grown on sapphire by vapour phase epitaxy", *Japanese Journal of Applied Physics,* no.32 , p. 1528-1533, 1993.

[HIR 02] HIRAMATSU K., NISHIYAMA K., MOTOGAITO A., MIYAKE H., IYECHIKA Y., MAEDA T., "Recent progress in the selective area growth and epitaxial lateral overgrowth of III-nitrides. Effect of reactor pressure in MOVPE growth", *Physica Status Solidi*, no.176, p. 535-543, 2002.

[KAG 06] KAGAMITANI Y., EHRENTRAUT D., YOSHIKAWA A., HOSHINO N., FUKUDA T., KAWABATA S., INABA K., "Ammonothermal epitaxy of thick GaN film using NH_4Cl mineralizer", *Japanese Journal of Applied Physics,* no. 45, p. 4018-4020, 2006.

[KEL 99] KELLY M.K., VAUDO R.P., PHANSE V.M., GORGENSE L., AMBACHER O., STUTZMANN M., "Large free-standing GaN substrates by HVPE and laser induced lift-off", *Japanese Journal of Applied Physics,* no. 38, L217, 1999.

[KID 00] KIDOGUCHI I., ISHIBASHI A., SUGAHARA G., BAN Y., "Air-bridged lateral epitaxial overgrowth of GaN thin films", *Applied Physics Letters,* no. 76, p. 3768, 2000.

[KIK 04] KIKUCHI A., KAWAI M., TADA M., KISHINO K., "InGaN/GaN multiple quantum disk nanocolumn light-emitting diodes grown on (111) Si substrates", *Japanese Journal of Applied Physics,* no. 43, p. L 1524–L 1526, 2004.

[KIM 04] KIM H.-M., CHO Y-H, LEE H., IL KIM S., RYU S.R., KIM D.Y., KANG T.W. and CHUNG K.S., "High-brightness LEDs using dislocation-free InGaN/GaN multiquantum-well nanorod arrays", *Nano Letters,* no.4, p. 1059-1062, 2004.

[KOU 97] KOUKITU A., TAKAHASHI N., SEKI H., "Thermodynamic study on metalorganic vapour phase epitaxial growth of group III nitrides", *Japanese Journal of Applied Physics,* no. 36, L1136, 1997.

[KRU 01] KRUKOWSKI S., BOCKOWSKI M., LUCZNIK B., GRZEGORY I., POROWSKI S., SUSKI T., WROBLEWSKI M., "High-nitrogen-pressure growth of GaN single crystals: doping and physical properties", *Journal of Physics: Condensed Matter,* no.13, p. 8881-8890, 2001.

[LAH 00] LAHRÈCHE H., LEROUX M., LAUGT M., VAILLE M., BEAUMONT B., GIBART P., "Buffer free direct growth of GaN on 6H–SiC by metalorganic vapor phase epitaxy", *Journal of Applied Physics,* no.87, p. 577-583, 2000.

[LEE 07] LEE S.W., MINEGISHI T., LEE W.H., GOTO H., LEE H.J., HANADA T., CHO M.W., YAO T., "Strain-free GaN thick films grown on single crystalline ZnO buffer layer with in situ lift-off technique", *Applied Physics Letters,* no.90, p. 061907, 2007.

[LER 97] LEROUX M., BEAUMONT B., GRANDJEAN N., LORENZINI P., HAFFOUZ S., VENNÉGUÈS P., MASSIES J., GIBART P., "Luminescence and reflectivity studies of undoped, n- and p-doped GaN on (0001) sapphire", *Materials Science and Engineering B*, no.50, p. 97-104, 1997.

[LES 95] LESTER S.D., PONCE F.A., CRAFORD M.G, STEIGERWALD D.A., "High dislocation densities in high efficiency GaN-based light-emitting diodes", *Applied Physics Letters*, no.66, p. 1249-1251, 1995.

[LIU 02] LIU L., EDGAR J.H., "Substrates for GaN epitaxy", *Materials Science and Engineering* R3, p. 61-127, 2002.

[LUR 86] LURYI S., SUHIR E., "New approach to the high quality epitaxial growth of lattice-mismatched materials", *Applied Physics Letters*, no.49, p. 140, 1986.

[MAN 72] MANASEVIT H.M., "The use of metalorganics in the preparation of semiconductor materials : Growth on insulating substrates", *Journal of Crystal Growth,* no.13-14, p. 306-314, 1972.

[MAT 01] MATHIS S.K., ROMANOV A.E., CHEN L.F., BELTZ G.E., POMPE W., SPECK J.S., "Modeling of threading dislocation reduction in growing GaN layers", *Journal of Crystal Growth*, no. 231, p. 371-390, 2001.

[MOT 02] MOTOKI K. , OKAHISA T., NAKAHATA S., MATSUMOTO N., KIMURA H., KASAI H., TAKEMOTO K., UEMATSU K., UENO M., KUMAGAI Y., KOUKITU A., SEKI H., "Growth and characterization of freestanding GaN substrates", *Journal of Crystal Growth,* p. 237-239 p. 912-921, 2005.

[NAG 00] NAGAHAMA S., IWASA N., SENOH M., MATSUSHITA T., SUGIMOTO Y., KIYOKU H., "High-power and long-lifetime InGaN multi-quantum-well laser diodes grown on low dislocation density GaN substrates", *Japanese Journal of Applied Physics,* no.39, L647-L650, 2000.

[NAK 91] NAKAMURA S., "GaN growth using GaN buffer layer", *Japanese Journal of Applied Physics,* no.30, L1705-L1707, 1991.

[NAK 94] NAKAMURA S., MUKAI T., SENOH M., "Candela-class high brightness InGaN/AlGaN double heterostructure blue-light-emitting diodes", *Applied Physics Letters,* no.64, p. 1687-1689, 1994.

[OSH 04] OSHIMA E., OGINO H., NIIKURA I., MAEDA K., SATO M., ITO M., FUKUDA T., "Growth of the 2-in-size bulk ZnO single crystals by the hydrothermal method", *Journal of Crystal Growth,* no. 260, p. 166-170, 2004.

[ROM 03] ROMANOV A.E., FINI P., SPECK J.S., "Modeling of the extended defect evolution in lateral epitaxial overgrowth of GaN: subgrain stability", *Journal of Applied Physics,* no.93, p. 106-114, 2003.

[SCH 99] SCHUBERT M., WOOLLAM J.A., LEIBIGER G., RHEINLANDER B., PIETZONKA I., SASS T., GOTTSCHALCH V., "Isotropic dielectric functions of highly disordered $Al_xGa_{1-x}InP$ ($0\leq x \leq 1$) lattice matched to GaAs", *Journal of Applied Physics,* no.86, p. 2025-2033, 1999.

[SHE 98] SHEN X.Q., TANAKA S., IWAI S., AOYAGI Y., "The formation of GaN dots on $Al_xGa_{1-x}N$ surfaces using Si in gas-source molecular beam epitaxy", *Applied Physics Letters,* no.72, p. 344-346, 1998.

[TAK 03] TAKEYA M., MIZUNO T., SASAKI T., IKEDA S., FUJIMOTO T., OHFUJI Y., OIKAWA K., YABUKI Y., UCHIDA S., IKEDA M., "Degradation of AlGaInN lasers", *Physica Status Solidi* (c) n°0 p. 2292-2295, 2003.

[TIN 05] TINJOD F., DE MIERRY P., LANCEFIELD D., BOUGRIOUA Z., LAÜGT S., TOTTEREAU O., LORENZINI P., CHENOT S. , VIREY E., KOKTA M.R., STONE-SUNDBERG J.L., PAUWELS D., J., "Alumina-rich spinel: a new substrate for the growth of high quality GaN-based light-emitting diodes", *Journal of Crystal Growth,* no. 85, p. 450-458, 2005.

[TSA 04] TSAO Y.J., "Lamps, chips, and materials for tomorrow", *IEE Circuits and Devices Magazine*, mai-juin 2004, p 28-37.

[VEN 00] VENNÉGUÈS P., BEAUMONT B., BOUSQUET V., VAILLE M., GIBART P., "Reduction mechanisms for defect densities in GaN using one- or two-step epitaxial lateral overgrowth methods", *Journal of Applied Physic,* no. 87, p. 4175-4181, 2000.

[VEZ 04] VÉZIAN S., NATALI F., SEMOND F., MASSIES J., "From spiral growth to kinetic roughening in molecular-beam epitaxy of GaN(0001)", *Physical Review B*, no. 69, p.125329-125 336, 2004.

[WAN 06] WANG G.T., TALIN A .A., WERDER D.J., RANDALL CREIGHTON J., LAI E., ANDERSON R. J., ARSLAN I., "Highly aligned, template-free growth and characterization of vertical GaN nanowires on sapphire by metal-organic chemical vapour deposition", *Nanotechnology,* no.17, p. 5773-5780, 2006.

[WU 98] WU X.H., FINI P., S.KELLER S., TARSA E.J., HEYING B., MISHRA U.K., DENBAARS S.P., SPECK J.S., "Dislocation generation in GaN heteroepitaxy", *Journal of Crystal Growth*, no.189-190, p. 231-245, 1998.

Chapter 3

III-Nitride High-Brightness Light-Emitting Diodes

3.1. Introduction

The mastery of GaN crystal growth, especially heteroepitaxial growth on sapphire substrates, was undoubtedly the key element to the development of high-brightness blue LEDs. The first attempts to synthesize GaN were made in the late 1960s [DIN 71] by using hydride vapor phase epitaxy (HVPE). This method led to excellent quality monocrystals but the samples were too size-reduced. For the following decade, the main technological obstacle indisputably remained the p-type doping of GaN, and this material was considered to be useless until the Professor Akasaki team research in 1988. Indeed, its work showed p-type doping under electron beam radiation of Mg-doped GaN layers [AMA 89]. At the same time, selenide II-VI semiconductors were sparking off great interest because of their wide bandgap suitable for blue-green light emission. Large industrial companies such as Sony, 3M or Philips aimed to fabricate short wavelength laser diodes in order to increase data storage capacities of CD-ROM or DVD optical readers/burners. Unfortunately, the lifetime of low threshold lasers developed in the mid 1990s could not exceed a few hundred hours, instead of the tens of thousands required for this kind of application. This limited lifetime can be explained by the increase in the number of non-radiative defaults under II-VI laser diode operating. Weak binding

---

Chapter written by Amélie DUSSAIGNE and Nicolas GRANDJEAN.

energy between atoms explains the low growth temperature used (300-500°C) and is responsible for structural default generation. This inherent characteristic makes II-VI-based semiconductors unsuitable for long lifetime optoelectronics applications. By way of comparison, temperatures used for GaN growth of optoelectronics devices (LEDs and lasers) are higher than 1,000°C.

Let us go back to the middle of the 1980s. As mentioned earlier, extensive research about II-VI semiconductors was carried out, whereas the scientific community neglected GaN-based compounds. Facing a fierce struggle between public and private laboratories for blue ZnSe laser market, a researcher called S. Nakamura working on the behalf of the Japanese company Nichia Chemical realized the necessity of tackling GaN field in order to ensure competitiveness. Nakamura thus went straight to his superiors and convinced them to provide $15 million for his research projects. He started by learning MOVPE growth at Florida University (USA) and then fabricated his own growth reactors (more details can be found in [NAK 00]). As such, Nakamura has accumulated an unsurpassed level of research in this field. He is currently professor at the University of California in Santa Barbara, where he still carries out GaN-based optoelectronics researches. Undoubtedly, this researcher had a great impact on the two last decades of III-nitride semiconductor research. We can particularly note the following findings: the discovery of the p-type doping by thermal annealing, the first bright blue light emitting diode [NAK 94], the first pulsed [NAK 96] and then continuous GaN laser, high luminous efficiency white LEDs [NAK 98], etc. Note that white LEDs are about to revolutionize general lighting by overriding other light sources such as neon tube thanks to the great advances in efficacy. Nowadays, white LEDs reach 160 lm/W, whereas neon tube and Edison's incandescent lamp efficacies are 80 and 15 lm/W respectively. The main obstacle that presently limits the broad penetration of white LEDs into the general lighting market is their prohibitive cost.

Although a high level of performances has been achieved by GaN-based optoelectronics devices, there is still a lot of room for improvement in different areas, especially since there are no GaN monocrystal substrates contrary to GaAs and Si. For example, GaN-based lasers only operate into a small wavelength domain centered around 405 nm. Their performances decay rapidly for long wavelength applications. To set an example, green lasers still do not exist today. This degradation trend is also relevant considering green LEDs, even though the latter are currently used for daily applications (traffic lights for instance). The efficiency of green LEDs is actually far lower than blue LEDs. The causes for this drop in efficiency will be discussed in detail later on in this chapter. They can partly be assigned to the internal electric field that arises from polarization effects present in the middle of the active region. On the other hand, they can also be explained by limitations due to "material" considerations while increasing indium content of the well.

Green LEDs constitute one of the actual key challenges for GaN-semiconductor domain since they could help in reaching, assuming efficiency close to blue LEDs, maximum light efficacy for white sources. This theoretical limit is set around 250 lm/W. A light device that combines a red, a green and a blue high-performance LED could reach light efficacy higher than 200 lm/W, i.e. twice the actual efficacy of neon tubes. Moreover, the luminous "comfort" could be unequalled since the color temperature would be close to the sun temperature. Finally, this kind of lighting system could allow the choice of the luminous atmosphere at the user's convenience. Thus, it is perfectly understandable that great lighting industrial companies (Osram, Philips) show great interest in general GaN LEDs research, and more particularly in green LED development.

This chapter will first address n- and p-type doping issues through the study of GaN p-n junction specificities. High-brightness active region LEDs that consists of one or several InGaN/GaN quantum wells will be discussed. Polarization field effects that appear in these heterostructures will then be studied. Finally, different parameters that control the radiative efficiency of the device will be tackled. The chapter will conclude with recent developments and future prospects that are emerging from the GaN LEDs field.

3.2. p-n junction in GaN

Doping GaN with acceptors to obtain a high concentration of holes has been a difficult problem. Initial Mg-doped layers grown by MOVPE were highly resistive because of hydrogen passivation effects on Mg-acceptors. The first p-n junction appeared only in the late 1980s, after the discovery that the Mg-doped GaN layer could be activated through electron beam exposure [AMA 89]. Although p-GaN doping is no longer a major obstacle in optoelectronic device fabrication, it is no less true that GaN Mg-doped layers account for a critical technological issue because of their still high resistivity. The main reasons are the deep level of Mg acceptor and the structural quality degradation of p-type epilayers when a high Mg-concentration is introduced. Be that as it may, Mg remains the best known p-type dopant for GaN-based semiconductors.

Let us begin with the description of the p-n junction theory applied to the particular case of nitrides. A p-n junction is composed of a n-type material (excess electrons) and a p-type material (excess holes). Ideally, all dopants are assumed to be fully ionized so that the free electron (holes) concentration is given by the donor (acceptor) concentration ($n=N_D$ and $p=N_A$). If $N_A=N_D$, the junction is symmetric and the current is principally due to diffusion effect (ideal junction). The recombination process then occurs both in the p-type GaN and in the n-type GaN. However, as discussed earlier, the main problem in a nitride-based p-n junction is the p-type doping. As Mg activation level is deep ($E_A \sim 200$ meV), all the p-dopant atoms cannot be ionized at 300 K ($N_A > p$). A high atomic concentration is thus required to obtain a p-type conductivity. On the other hand, Mg atom excess can lead to an auto-compensation effect. The material then obtains insulating properties because of the simultaneous formation of nitrogen vacancies (V_N) and/or MgV_N (Mg_2V_N) complexes. While using the MOVPE technique, this phenomenon typically appears for Mg atomic concentrations higher than 2.10^{19} at/cm^3. In comparison with the n-type GaN, the ionization energy of Si atoms is weak ($E_D \sim 30$ meV @ 300 K), so that the number of free electrons is equal to the number of donors ($n \sim N_D$).

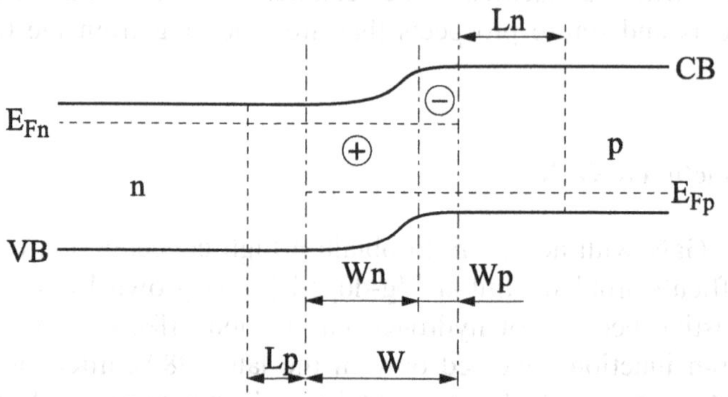

Figure 3.1. *p-n junction band diagram at thermal equilibrium. E_{Fn} (E_{Fp}) is the quasi-Fermi level in the p (n)-type region. CB is the conduction band and VB is the valence band. W is the depletion layer width, $W = W_n + W_p$. L_n (L_p) is the diffusion length of electrons (holes)*

An electric field resulting from the presence of + and – charges appears in the p-n junction. Indeed, the neutrality condition is lost after the free carrier diffusion and the only charge in the depletion region is from ionized donors and acceptors. This depleted region, also called the space charge region (SCR), is characterized by its width W (W_n on the n-type side and W_p on the p-type side). It presents + charges on the n-type side and – charges on the p-type side. This is clearly illustrated in Figure 3.1. We can notice that the depletion region extends more in the less doped region. Let us recall here that the parameter taken into account in the depletion layer width calculation is the number of donor/acceptor atoms, and not the electron/hole concentration.

As previously discussed, $N_A > N_D$ in the particular case of a GaN p-n junction since a high number of Mg atoms is needed to reach a reasonable conductivity (because of the deepness of the acceptor level). Only a few percents of Mg atoms are ionized at 300 K. Thus, the SCR extends mainly in the n-type region. The extension lengths of the space charge region in GaN p- and n-type of an unbiased junction are given by:

$$W_n^2 = 2\varepsilon V_d/(eN_d(1+N_d/N_a)) \quad [3.1]$$

$$W_p^2 = 2\varepsilon V_d/(eN_a(1+N_a/N_d)) \quad [3.2]$$

where Vd is the diffusion voltage, e is the elementary charge and ε is the vacuum permittivity. The diffusion voltage is set as the potential difference between the neutral regions of opposite conductivity type. It can be written:

$$V_d = (kT/e)\ln(N_dN_a/n_i^2) \quad [3.3]$$

where n_i is the intrinsic carrier concentration. Typically, the electron concentration is of the order of $3\text{x}10^{18}$ cm^{-3} considering an equivalent Si atom concentration, whereas the hole concentration is of the order of $5\text{x}10^{17}$ cm^{-3} for a Mg atomic concentration higher than $1\text{x}10^{19}$ cm^{-3}. It roughly corresponds to an SCR width of 29 nm on the n-type side and 9 nm on the p-type side, under unbiased conditions. As a result, the higher the Mg doping level, the more shapeless the junction is. Furthermore, hole injection is difficult because of the low hole mobility ($\mu_p \sim 10$ cm²/V/s) compared to the electron mobility ($\mu_n \sim 200$ cm²/V/s). Consequently, under low current injection, the p-type region controls the current flow. It raises the point that if a standard GaN p-n junction is used, most of the recombinations will occur into the p-type region. We could thus consider growing the active layer, which consists of InGaN/GaN quantum wells, into the p-type region. Unfortunately, a strong impurity concentration could lead to critical phenomena of InGaN alloy inter-diffusion that can affect radiative efficiency.

In order to prevent recombination mechanisms from taking place in the p-type GaN, an electron diffusion barrier known as the "electron blocking layer" is introduced. This technique is currently used in GaN LEDs [NAK 94] and even in some cases of LEDs based on arsenide III-V compounds. Figure 3.2 displays the band diagram of such a junction under forward bias conditions. Considering the conduction band, the electron barrier is too high so that they cannot pass over it and stay localized on the n-side of the junction. On the other hand, the hole barrier in the valence band is lower because of the smaller bandshift and the Mg-doping. Holes can then pass over quite easily. Consequently, recombinations take place either in the n-type GaN or in the SCR.

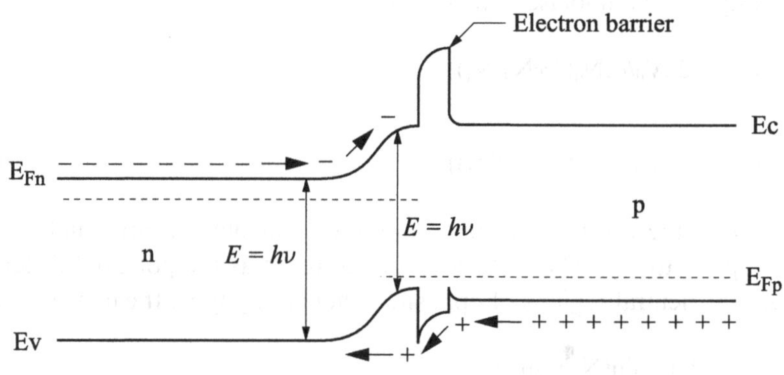

Figure 3.2. *Band diagram of an AlGaN electron blocking layer p-n junction under forward bias*

3.3. Active region: InGaN/GaN quantum well

In order to increase the recombination probability in the active region, it is common practice to insert quantum wells, i.e. material layers of smaller bandgap. Its thickness has to be of the order of the characteristic wavelength associated with the electron. As carrier effective mass is very high in the case of III-N compounds, the typical quantum well thickness that allows quantum confinement is very small, in the nanometer scale. What is more, as will be explained later on, heterostructures are epitaxied along a polar axis ((0001) GaN axis in the general case of growth on sapphire substrate). In this configuration, heterostructures present a strong internal electrical field that leads to a giant quantum confined Stark effect. Now, this phenomenon is critical for the internal efficiency of LEDs, because it strongly decreases the recombination rate in quantum wells by spatially separating electrons from holes. Consequently, quantum well thickness of high-brightness LEDs should be kept very thin in order to limit this separation.

3.3.1. *Growth and structure*

GaN lattice fitting substrates do not exist, except GaN itself, but its fabrication is extremely critical and expensive. What is more, GaN substrate size does not exceed 1 cm^2. Though the sapphire substrate presents a strong mismatch with the GaN lattice parameter (-33% without epitaxial relation, 16% with epitaxial relation), it finally established itself as the most extensively used substrate for LED fabrication owing to its thermal dilatation coefficient, which is close to that of the GaN, as well as its low cost.

However, the strain resulting from GaN heteroepitaxy on sapphire leads to the formation of dislocations which propagate through the GaN layer. Considering MBE growth, the dislocation density obtained is around $5x10^9/cm^2$. In MOVPE growth, thanks to different growth processes (epitaxial lateral growth for instance), it can be lowered until $10^7/cm^2$. It reaches $10^6/cm^2$ and even less in very thick layers grown by HVPE.

Table 3.1 gives lattice parameters of commonly used nitride compounds in their hexagonal geometry. As the active region is usually formed by (Ga,In) N/GaN multiple quantum wells, we will pay particular attention to $In_xGa_{1-x}N$ growth over GaN. The lattice mismatch between GaN and InN is 11%.

We will further see that this important value can lead to issues during In incorporation into the GaN matrix.

Materials	**a (Å)**	**Δa/a (%)**
GaN	**3.189 [1]**	**-**
AlN	**3.113 [1]**	**-2.4**
InN	**3.538 [2]**	**10.9**
Sapphire	**4.758**	**16.1**

Table 3.1. *Lattice parameters (a) and lattice mismatches of GaN with AlN, InN and sapphire substrate*

The lattice mismatch between GaN and InN is given by the following formula:

$$\Delta a/a = (a_{InN} - a_{GaN})/a_{GaN} \qquad [3.4]$$

where a_{InN} and a_{GaN} are InN and GaN lattice parameters respectively. The Vegard's law makes it possible to directly calculate the lattice parameter of the $In_xGa_{1-x}N$ alloy:

$$a_{InGaN} = x\ a_{InN} + (1 - x)\ a_{GaN} \qquad [3.5]$$

If the InGaN alloy accommodates 20% In, the lattice mismatch with GaN is 2.2%. Consequently, InGaN layers are strongly strained over GaN and dislocations can appear upon a certain critical thickness (t_c). Of course, t_c varies with the In content. In the particular case of GaN LED quantum wells, the InGaN layer thickness is generally smaller than 3 nm for an In content of less than 25%. In these conditions, structures can be considered as pseudomorphic.

At thermodynamics equilibrium, the In miscibility level into GaN is low because of a weak binding energy between In and N as well as an important lattice mismatch of GaN with InN. Consequently, InGaN alloy growth over GaN has to be done far away from the thermodynamics equilibrium, in order to obtain a good material quality from the "optoelectronics" perspective. In practice, the temperature is lowered by several hundred degrees comparing to the standard GaN growth temperature. For instance, the temperature drops from 1,000-1,100°C for

GaN to 700-800°C for InGaN during a MOVPE process. The maximal In composition is situated around 20-25% to ensure a sufficient quality to high-brightness LEDs. We will see that this limited composition of InGaN alloys is the actual technical obstacle to the fabrication of highly efficient green light emitting diodes.

Phase segregation of InGaN is commonly observed in (Ga,In)N/GaN quantums wells and especially in those grown by MOVPE. These composition variations have been identified as the origin of exciton localization which leads to the extraordinary efficiency of nitride LEDs considering the highly defective properties of GaN. Other research groups reported quantum wells with In-rich phases, composition inhomogenity, agregats and even nano-objects similar to InN pure quantum dots. However, these results have to be considered cautiously because they rely on transmission electron microscopy (TEM) experiments. Indeed, it has recently been demonstrated that the electron beam used during the observation could be responsible for In clustering. The end of this chapter will be dedicated to the role played by the In composition fluctuations in the outstanding radiative efficiency of LEDs based on III-nitride semiconductors.

3.3.2. *Optical properties*

3.3.2.1. *Quantum wells and radiative efficiency*

Figure 3.3 shows the emission wavelength versus photoluminescence intensity of InGaN/GaN quantum wells grown by EPFL (Ecole Polytechnique Fédérale de Lausanne). This spectrum reflects the internal quantum efficiency of the quantum wells according to the wavelength. Note that quantum external efficiency of LEDs produced by Nishia, the pioneer and still the leader of the domain, follows the same trend [NAK 99]. Internal quantum efficiency increases from 370 nm to 450 nm and then drops beyond 450 nm.

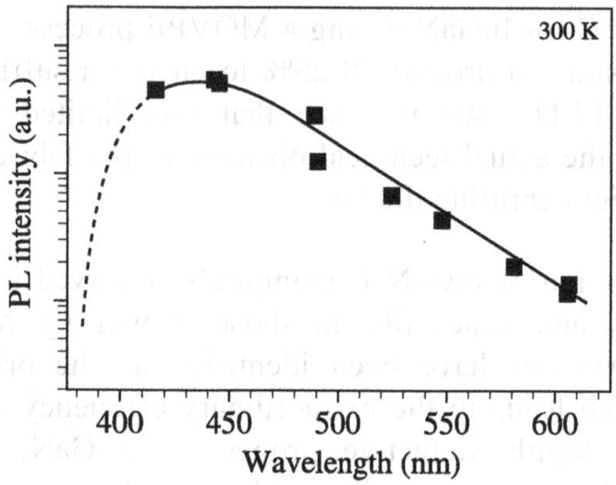

Figure 3.3. *PL intensity versus emission wavelength of InGaN/GaN quantum wells*

Low efficiency at short wavelength (<400 nm) can be assigned to different phenomena. On the one hand, the lower well width and In content can decrease potential barriers in the conduction and in the valence band and thus facilitate the thermal escape of carriers. On the other hand, a lower In composition decreases the composition fluctuation effect which presumably explains high radiative efficiency despite the high density of dislocations. We will come back on this point during the discussion about LED efficiency. Least but not last, photoluminescence signal re-absorption has to be taken into account since barriers are no longer completely transparent as we get closer to the GaN gap (362 nm @ 300 K).

In the long wavelength domain (>500 nm), the radiative efficiency drop cannot be explained by absorption phenomena or by thermal escape, since In composition and well width are both high. How can we explain the external quantum efficiency drop for wavelengths longer than 500 nm?

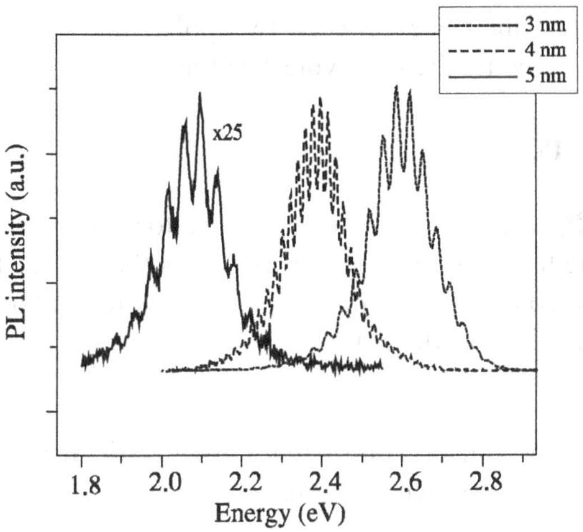

Figure 3.4. *PL spectra of In0.22Ga0.78N/GaN quantum wells emitting from blue to orange @ 300K (see color plate section)*

Figure 3.4 presents PL spectra of three InGaN/GaN wells, whose emission wavelength is situated between 450 and 600 nm according to their thickness (3, 4 and 5 nm). We can notice a dramatic drop of PL signal for longer wavelength. As only the well thickness varies in this experiment, the degradation cannot be attributed to a poorer quality of InGaN alloy. On the other hand, it could be possible that the critical thickness of plastic relaxation has been reached, leading to the appearance of structural defects, but such is not the case here. The explanation is elsewhere and is linked to the presence of an internal electric field into nitride heterostructures [BER 97, LER 98, BUT 08].

3.3.2.2. *Quantum confined Stark effect*

Polarization phenomena and internal electric field

Wurtzite symmetry implies that positive and negative centers of mass arising from atoms binding do not coincide. This leads to a spontaneous macroscopic polarization across the crystal [BER 97]. Besides, InGaN layers grown over GaN are strained since there is a lattice mismatch between the two materials. This strain also results in a polarization effect characterized as "piezoelectric". We will relate the internal electric field present in a quantum well to both the spontaneous and piezoelectric polarizations which appear in the materials.

The SCR at the interface between two polar materials that constitute a quantum well is given by the following relation:

$$\sigma_{intb/w} = \mathbf{n}.(\mathbf{P_b} - \mathbf{P_w}) \qquad [3.6]$$

where $\mathbf{P_b}$ and $\mathbf{P_w}$ are the total polarizations for the barrier and the well respectively. The total polarization for its part is the sum of the spontaneous and piezoelectric components. The vector unit $\mathbf{n}$ is normal to the surface and directed according to the (0001) axis, when GaN is grown on the c-plane sapphire.

If we consider the conservation law for the electric displacement $\mathbf{D} = \mathbf{F} + \varepsilon\mathbf{P}$ at the barrier/well interface, we obtain:

$$\varepsilon_w\varepsilon_0\mathbf{F_p} - \varepsilon_b\varepsilon_0\mathbf{F_b} = \mathbf{P_b} - \mathbf{P_p} \qquad [3.7]$$

where ε_0 is the free space permittivity, and ε_w and ε_b are the dielectric constants of the well and the barrier respectively. In the particular case of infinite barriers, the fields within the barriers due to interface charges cancel each other out. The field within the well can thus be simply written:

$$\mathbf{F_w} = (\mathbf{P_b} - \mathbf{P_w}) / \varepsilon_w\varepsilon_0 \qquad [3.8]$$

If the barriers are thin, while considering multiple quantum wells for instance, the electric field spreads out along the structure. This can be taken into account by adding a form factor. Each period (well + barrier) introduces a potential drop equal to $-e(L_bF_p+L_wF_w)$. L_b and L_w are the barrier and well thicknesses. By setting linking conditions at the structure extremities, the following relation can be obtained: $L_bF_b + L_wF_w = 0$. While injecting this condition within the electrical displacement vector relation, we obtain:

$$\mathbf{F}_w = L_b(\mathbf{P}_b - \mathbf{P}_w)/(\varepsilon_w\varepsilon_0 L_b + \varepsilon_b\varepsilon_0 L_w) \qquad [3.9]$$

Assuming well and barrier dielectric constants are equal, then:

$$\mathbf{F_w} = L_b/[L_b + L_w](\mathbf{P_b} - \mathbf{P_w})/\varepsilon\varepsilon_0 \qquad [3.10]$$

Note that in the case of infinite barriers, the electric field equation for a single quantum well is conserved.

In (Ga,In)N/GaN quantum wells, the spontaneous polarization difference between GaN and $In_xGa_{1-x}N$ can be neglected, considering the piezoelectric polarization. Consequently, the electric field present into the well is only dependent on the well. Typically, the experimental value of the electric field into (Ga,In)N/GaN quantum wells that contains 20% In is of the order of 2.5 MV/cm [LEF 01].

The theoretical value calculated by Bernardini *et al.* is 3.3 MV/cm considering the same In composition [BER 01].

This shift can undoubtedly be explained by the non-ideal character of the structure, especially for $In_xGa_{1-x}N$. Actually, concentration fluctuations are important during the growth of this alloy.

Quantum confined Stark effect

This internal electric field leads to the quantum confined Stark effect (QCSE) which shifts optical transitions towards lower energies [BUT 08]. Ground state transition $E_{e1\text{-}hh1}$ can thus be written:

$$E_{e1\text{-}hh1} = e_1 + hh_1 + E_g - E_{Ry} - eF_pL_p \qquad [3.11]$$

where e_1 and hh_1 are the electron and hole fundamental energy levels in the quantum well, E_g is the well bandgap, E_{Ry} is the excitonic binding energy, F_p is the electric field present into the well and L_p is the well thickness. The last term in this relation comes from the band bending due to the electric field and it explains the transition redshift (Figure 3.5).

Let us recall this electric field is huge in nitrides. As an example, its value reaches 2.5 MV/cm in the $In_{0.2}Ga_{0.8}N$/GaN quantum well. Consequently, if the well thickness increases by 1 nm, the transition energy decreases by 250 meV. It represents a wavelength fluctuation of a 20 nm. It is a very important feature of the InGaN material system since, thanks to this effect, nitride-based LEDs can reach the red spectral domain.

Indeed, it has previously been mentioned that InGaN alloy quality deteriorates beyond an In content of 20 to 25%.

This composition sets the upper wavelength attainable by a bulk layer, i.e. 470 nm. However, thanks to the QCSE, it is possible to obtain emission in the red wavelength range by simply increasing the well thickness.

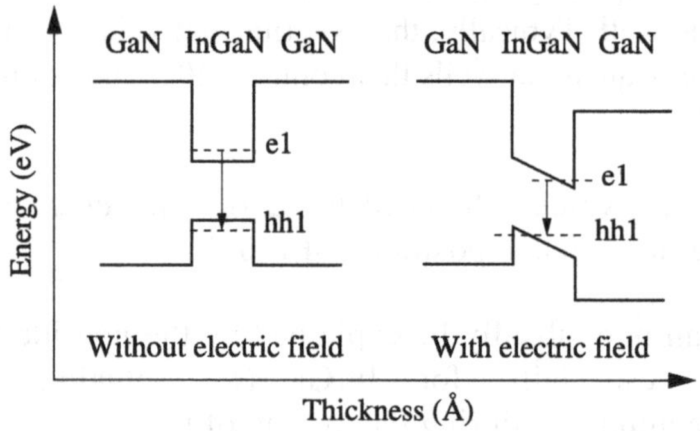

Figure 3.5. *Potential profile of quantum wells with and without internal electric field*

Unfortunately, the price of increasing the well thickness is a sharp drop of radiative efficiency due to spatial separation of electron-hole pairs. Oscillator strength is proportional to the square of the overlap integral of wave functions and is all the more weak that electron and hole wave functions are spacially separated (Figure 3.6a). A quick calculation shows that QCSE becomes significant on oscillator strength when wells are thicker than 3 nm (Figure 3.6b).

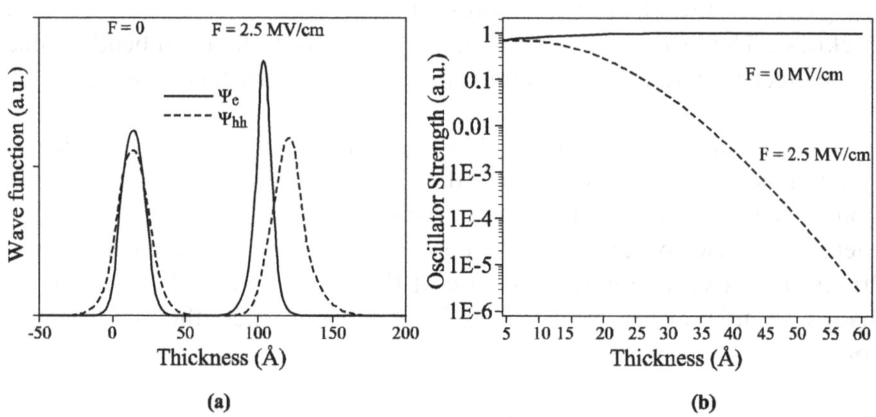

Figure 3.6. *(a) Electron and hole wave functions in a quantum well with and without internal electric field and (b) oscillator strength versus well thickness*

Thus, for a given well thickness, the stronger the electric field, the more the oscillator strength is reduced. In the same manner, the thicker the well, the weaker the oscillator strength is for a given In content. To reach low energies (550 nm for example), it is necessary to have the maximum electric field, i.e. around 20% In (F= 2.5 MV/cm). Well thickness will then be fixed at 3.2 nm instead of 1.6 nm in the case of a 450 nm wavelength. Oscillator strength is then 16 times reduced between 450 and 550 nm. Let us mention that in the case of a 1.6 nm thick well, the confinement effect is stronger than the QCSE and the oscillator strength is only slightly affected by the internal electric field. The reachable energies according to the well thickness and composition are illustrated in Figure 3.7.

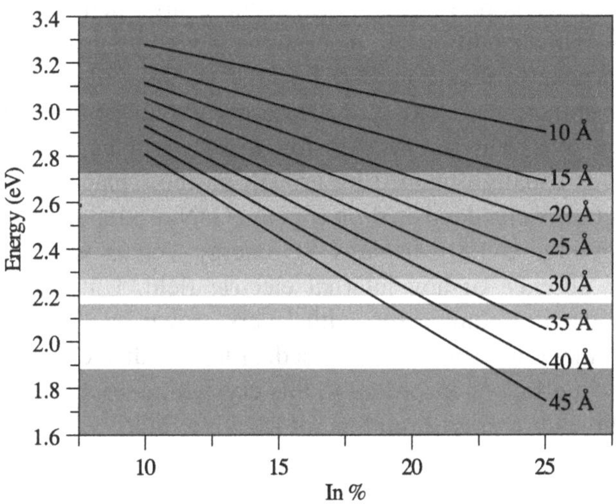

Figure 3.7. *InGaN/GaN quantum well transition energy considering both In% and well thickness (see color plate section)*

As previously discussed, QW thickness should not pass over 3 nm in order to limit the QCSE and retain a suitable radiative efficiency. Long wavelength LEDs thus require an In composition increase in order to keep their radiative efficiency constant. However, InGaN alloy quality quickly deteriorates over 20-25% In. Consequently, green LEDs contain thicker wells, which is why their performances are dramatically reduced. Commercial blue LEDs use InGaN layers with 15-20% In and wells thinner than 3 nm. This type of structure, assuming strong efforts on light extraction, is able to reach at least 50% external quantum efficiency.

It is clearly underlined in the previous explanation that QCSE is a critical obstacle to long wavelength emitter fabrication by using thick quantum wells in which the quantum efficiency is strongly reduced. This is a reason why green LEDs future achievements depend on new developments that aim to suppress the internal polarization field.

3.3.2.3. *Polar and non-polar orientations*

It has been shown in the previous section that the electric charge induced by the macroscopic polarization deviation between two materials is dependent on the angle that appears between the vector normal to the interface and the vector related to the macroscopic polarization. In other words, if the latter is perpendicular, then the charge density is going to be equal to zero. This condition can be easily achieved by growing the material along a non-polar orientation [WAL 00]. As the polarization vector is directed according to the c-axis, it is necessary to choose a growth plane that contains the c-axis. This is the case for *a*- and *m*-plane, for example. The GaN *a*-plane is usually obtained by growth on the *r*-plane sapphire (Figure 3.8a). 3 quantum wells deposited on *a*-plane GaN are reported in Figure 3.8. The curve in Figure 3.8b displays PL energy versus well thickness and confirms the absence of any internal electric field. Unfortunately, crystal quality is reduced in comparison with layers grown on *c*-plane growth and various explanations can be advanced. Firstly, dislocation annihilation mechanism is inefficient according to this crystalline orientation. Then, GaN *a*-axis growth leads to a great number of stacking faults.

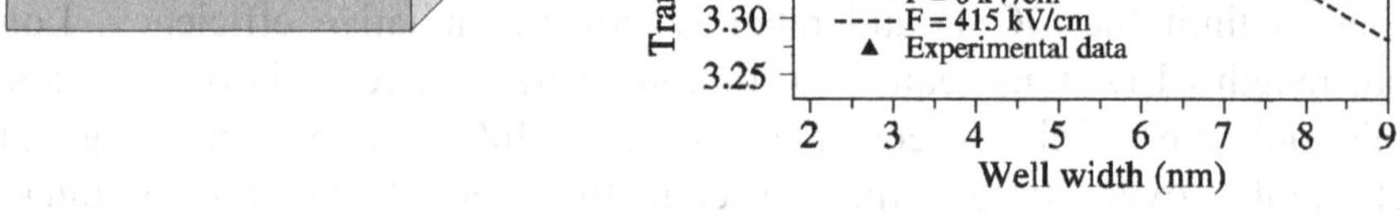

(a) (b)

Figure 3.8. *(a) Schematic view of an a-plane GaN layer grown on r-plane sapphire substrate and(b) transition energy of GaN/AlGaN QWs grown on a-plane*

Nowadays, research efforts turn towards GaN m-axis growth because of a better surface state, but the critical issue remains the substrate. First results obtained on m-axis oriented GaN substrates are very promising [KIM 07]. Unfortunately, these substrates are still size-reduced since they originate from GaN quasi-substrates (or bulk substrates) which are c-axis grown. *m*-plane substrates are consequently only obtained after lateral cutting.

Another way to proceed in order to cancel the internal electric field consists of growing GaN heterostructures on a semi-polar surface [TAK 00, GIL 07]. In this case, it is possible to take advantage of the fact that the internal electric field is given by the total polarization difference between the two materials. This difference, which depends on piezoelectric and spontaneous components, can be cancelled according to the surface orientation, but here again, the substrate issue remains unsolved.

The main motivation which leads world research towards "non-polar" and "quasi-polar" surfaces lies in their strong potential for green light emission. As InGaN alloy quality deteriorates very quickly while increasing In composition, long wavelengths are thus reachable only by increasing QW thickness. However, we have seen that LED efficiency is strongly affected by the dramatic drop of oscillator strength due to the electric field present in heterostructures grown along the c-axis. Consequently, the absence of an electric field on *a*- and *m*-planes should help in reaching higher efficiency in the green light domain. Nevertheless, to fulfill such an aim, InGaN alloys composed of more than 30% In are required. Much research is still needed in the materials science field, though the last advances are very promising since recent non-polar LED efficiency exceeds 30% [KIM 07].

3.4. Radiative efficiency

One of the most spectacular particularities of InGaN QW-based LEDs is inherent to their internal quantum efficiency (from 70-80%). These values are absolutely exceptional considering the poor current epilayer quality. Indeed, large area bulk GaN substrates are not commercially available. Then, sapphire became the most extensively used substrate for growth of GaN crystals. The lattice mismatch of GaN with sapphire is around 33% (16% considering the epitaxial relation that generates a 30° GaN unit cell rotation with regards to sapphire). Be that as it may, such a shift in lattice parameters leads to plastic relaxation of strained layers. A great number of dislocations are thus created. According to their characteristic (edge or screw), the dislocations only partly cancel each other out with layer thickness. As a result, the typical density of threading dislocations, i.e. those that pass through the InGaN/GaN epitaxial films and emerge at the surface, is at least 10^8 cm^{-2}. It is interesting to note that first Nishia LEDs exhibited an even higher dislocation density (10^9 to 10^{10} cm^{-2}). Despite this huge value, the devices showed very high level performances.

Why do highly defective GaN-based materials emit brilliant light? It is certain that any other III-V optoelectronic devices, based on GaAs for instance, could not work under these conditions. Thus, one of the first explanations relying on experimental results tried to prove the inactive character of these dislocations. However, it has been quickly experimentally demonstrated that GaN dislocations are actually non-radiative recombination centers that should quench the light emission.

Narukawa *et al.* [NAR 97] put forward another explanation based on Nichia Chemical sample TEM observations. It appeared on TEM images that InGaN/GaN quantum wells looked more like quantum boxes, probably because of phase separation of the InGaN alloy. This research group thus proposed that carriers (electrons and holes) are strongly localized into In-rich clusters. This localization prevents carrier diffusion towards dislocations and limit non-radiative recombinations (Figure 3.9). It is an interesting point to note that the idea of using quantum boxes in a default rich material has been previously expounded by J.M. Gérard and C. Weisbuch [GER 91]. This idea was even demonstrated in 1996 in the case of InA/GaA quantum boxes on the silicon substrate [GER 96].

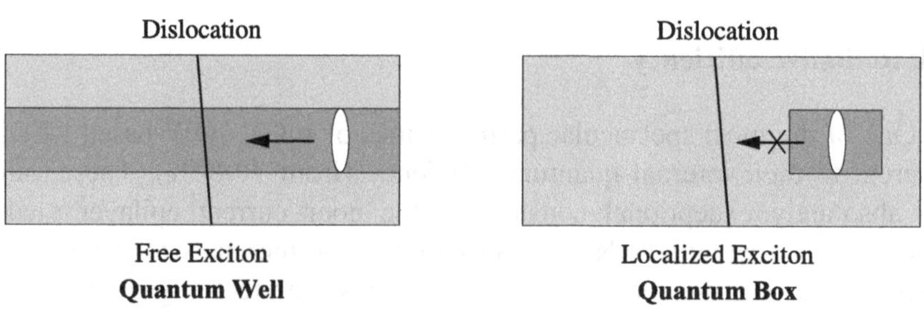

Figure 3.9. *Illustration of carrier localization into a quantum box system*

Be that as it may, the outstanding quantum efficiency of blue LEDs has been commonly explained for many years by the presence of these nanometer-scale In-rich clusters. However, in 2003, C. Humphreys from the University of Cambridge brought this model into question. Indeed, he proved with supporting videos that the In-rich regions were actually created under electron beam radiation during TEM observation [SME 03]! After one year full of debate and controversy, TEM observation artefacts have finally been admitted. However, there was still no acceptable explanation considering LED efficiency. In 2005, the A. Hangleitter group suggested a new mechanism based on both TEM and AFM (atomic force microscopy) observations. Threading dislocations generate depressions at the growth surface in the form of reverse hexagonal pyramids [HAN 05]. Consequently, epitaxied quantum wells present a topographical default since they become thinner on the depression side (Figure 3.10). What is more, parts of the quantum well are now tilted with regards to the polar plane (0001). This leads to a decrease of the internal electric field and consequently an attenuation of the QCSE. As a consequence of these two effects, the transition energy increases for the parts of the quantum well situated on the depression sides that surround the dislocation. In other words, dislocations are surrounded by a potential barrier which prevents carrier diffusion and thus non-radiative recombinations.

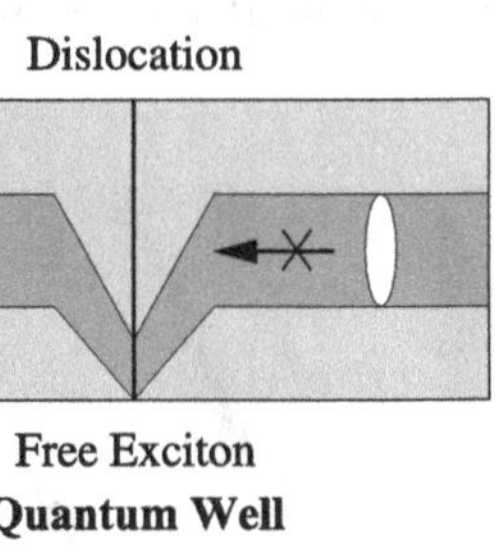

Figure 3.10. *Screening effect illustration of a dislocation surrounded by a depression (from [HAN 05])*

This model appears very attractive, but unfortunately, commercial high-brightness LEDs do not seem to present these kinds of defects. More recently, other assumptions have been expressed. As an example, it has been proposed that In-atom clusters present in InGaN could be responsible for hole localization into the valence band [CHI 06]. Another study [GRA 05] suggests that quantum well thickness variation corresponding to a single monolayer would be sufficient to localize excitons since QCSE leads to an energy fluctuation $\Delta E = F \times \Delta Lw$, where F is the internal electric field and Lw is the quantum well width. If we consider F to be around 2 MV/cm (value commonly reported for 15% In) and a monolayer fluctuation (0.26 nm), it results that $\Delta E \approx 50$ meV. However, this value remains low to generate a strong confinement of the carriers. In 2006, our team proposed an alternative explanation based on an AFM extensive study of the InGaN/GaN quantum well surface. AFM observations show a surface morphology characterized by very marked valleys (Figure 3.11b). The topographical profile confirms that the quantum well thickness varies considerably and can almost be zero in these valleys. In other words, the quantum well energy transition strongly increases at the vicinity of the valleys. This point has been confirmed by time resolved cathodoluminescence measurements. We can clearly see that zones revealed as very thin by AFM correspond to higher transition energies (Figure 3.11a) with regards to nominal energies (Figure 3.11.c). According to our interest, the important item is that dislocations seem to be at the root of valley formation and are thus situated at their bottom. Consequently, dislocations are screened by a potential barrier which comes from the thickness decrease of the quantum well at their proximity. Carriers can on longer recombine non-radiatively on the dislocations. This mechanism is similar to that proposed by Hangleitter *et al.* [HAN 05], though the origin of the potential barrier formation around dislocations is different.

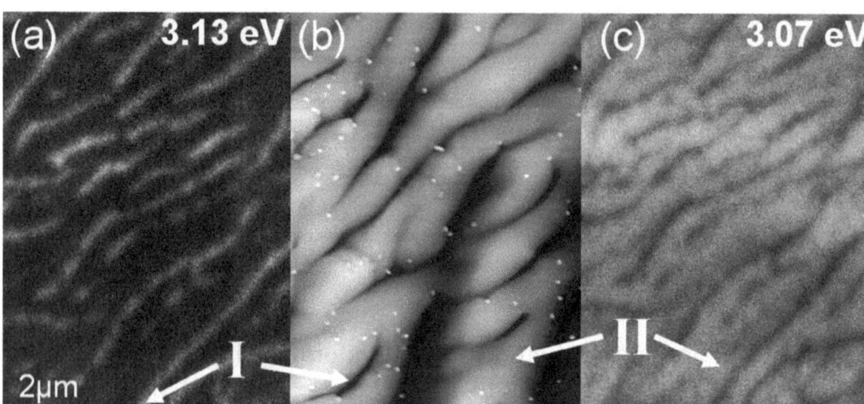

Figure 3.11. *(a) CL images of the quantum well for high energy, (b) AFM image of a thin InGaN layer that corresponds to a typical quantum well thickness and (c) luminescence nominal energy*

Be that as it may, the issue of radiative efficiency of InGaN/GaN QW-based LEDs is still being discussed. The point which has been highlighted through the different studies is that exciton localization is required to prevent non-radiative recombination on dislocations. Let us note that the growth technology that allows a high radiative efficiency is the MOVPE technique. Hence, all the studies discussed in the previous section refer to samples grown using this technique. To date, MBE growth technology has not achieved comparable results. It can be suspected that growth conditions which are further from the thermodynamics equilibrium limit the previously discussed effects, i.e. potential barrier formation around dislocations. An advantageous approach would be to substitute the well by quantum boxes aligned on a plane. Unfortunately, the first results published do not present convincing findings.

3.5. Conclusion and prospects

InGaN/GaN QW-based blue LEDs grown on c-plane sapphire substrate have reached an advanced degree of maturity and a high level of performance, i.e. external quantum efficiency higher than 50%. This is all the more outstanding as the active layer is crossed by a great number of threading dislocations ($>10^8$ cm^{-2}). There is no other kind of system able to achieve such performance regarding such a poor crystal quality. The ability to be insensitive to defaults is undoubtedly the main feature of nitride-based LEDs.

If blue LEDs have probably reached their limits in term of efficiency, the situation is different considering green and even shorter wavelength devices. This limit comes from the internal electric field that consists of a piezoelectric and a spontaneous component. This phenomenon dramatically decreases radiative efficiency of thick wells. However, let us recall that large wells are necessary to counterbalance the poor crystalline quality of In-rich InGaN alloys. Thus, the up and coming solution is the growth on non-polar or quasi-polar faces. In this case, the internal electric field is cancelled and thicker wells are consequently allowed. Nevertheless, 30 to 35% In is required to reach the green spectral domain (520 to 530 nm). It constitutes a real challenge and extensive work is still needed to carry out green LED performances on the scales of blue LEDs. The last critical issue that needs to be highlighted is the growth on non-polar or quasi-polar substrates. The best results have been obtained with samples grown on *m*-plane quasimonocrystalline GaN substrates. However, the main disadvantage remains the size, since these substrates do not reach 1 cm^2. Many improvements can also be achieved in this field and future advances will be decisive to establish an alternative technology to the conventional growth on c-plane sapphire substrates.

Nowadays, the majority of world research efforts are spent on the pursuit of the green spectral domain. However, it is important to keep in mind that blue LEDs (460-470 nm) presently offer outstanding performances. This high level of achievement led to the fabrication of white LEDs which outperform fluorescent tubes in terms of luminous efficiency. In figures, the luminous efficiency of white LEDs reached 170 lm/W at the end of 2007. This value was unforeseen a few years ago. Of course, we are still far from the common use of LEDs in general lighting, but the main technological barriers have been removed. Then, it will be difficult to stop this technological revolution and GaN will become to the optoelectronics domain what Si presently represents for the electronics field.

3.6. Bibliography

[AMA 89] AMAMO H., KITO M., HIRAMATSU K., AKASAKI I., "P-type conduction in Mg-doped GaN treated with low-energy electron beam irradiation (LEEBI)", *Japanese Journal of Applied Physics*, no. 28, L2112, 1989.

[BER 01] BERNARDINI F. and FIORENTINI V., "*Nonlinear macroscopic polarization in III-V nitride alloys*", *Physical Review B,* no.64, p. 085207, 2001.

[BER 97] BERNARDINI F., FIORENTINI V., VANDERBILT D., "Spontaneous polarization and piezoelectric constants of III-V nitrides", *Physical Review B,* no.56, R10024-R10027, 1997.

[BUT 08] BUTTÉ R., GRANDJEAN N., "Effects of polarization in Optoelectronic quantum structures", in C. Wood and D. Jena (eds.), *Polarisation Effects in Semiconductors*, Springer, Berlin, 2008.

[CHI 06] CHICHIBU S.F., UEDONO A., ONUMA T., HASKELL B.A., CHAKRABORTY A., KOYAMA T., FINI P.T., KELLER S., DENBAARS S.P., SPECK J.S., MISHRA U.K., NAKAMURA S., YAMAGUCHI S., KAMIYAMA S., AMANO H., AKASAKI I., HAN H.J., SOTA T., "Origin of defect-insensitive emission probability in In-containing (Al,In,Ga)N alloy semiconductors", *Nature Materials* no.5, p. 810-816, 2006.

[CHI 96] CHICHIBU S., AZUHATA T., SOTA T., NAKAMURA S., "Spontaneous emission of localized excitons in InGaN single and multiquantum well structures", *Applied Physics Letters*, no.69, p. 4188-4190, 1996.

[DIN 71] DINGLE R., SELL D.D., STOKOWSKI S.E., ILEGEMS M., "Absorption, reflectance and luminescence of GaN epitaxial layers", *Physical Review B,* no.4, p. 1211, 1971.

[GER 91] GÉRARD J.M., WEISBUCH C., "Semiconductor structure for optoelectronic devices", French patent no. 91400015.3, 1991.

[GER 96] GÉRARD J.M., CABROL O., SERMAGE B., "InAs quantum boxes: highly efficient radiative traps for light emitting devices on Si", *Applied Physics Letters,* no.68, p. 3123, 1996.

[GIL 07] GIL B., "Symmetry properties and anisotropic excitonic response for GaN films grown on semipolar (11–22)-, (10–11)-, and (10–13)-oriented GaN substrates", *Applied Physics Letters,* no.90, p.121903, 2007.

[GRA 05] GRAHAM D.M., SOLTANI-VALA A., DAWSON P., GODFREY M.J., SMEETON T.M., BARNARD J.S., KAPPERS M.J., HUMPHREYS C.J., THRUSH E.J., "Optical and microstructural studies of InGaN/GaN single-quantum-well structures", *Journal of Applied Physics,* no.97, p. 103508-1035012, 2005.

[HAN 05] HANGLEITER A., HITZEL F., NETZEL C., FUHRMANN D., ROSSOW U., ADE G., HINZE P., "Suppression of non-radiative recombination by V-shaped pits in GaInN/GaN quantum wells produces a large increase in the light emission efficiency", *Physical Review Letters,* no.95, p. 127402, 2005.

[KIM 07] KIM K.C., SCHMIDT MC., SATO A.H., WU F., FELLOWS N., JIA Z., SAITO M., NAKAMURA S., DENBAARS S.P., SPECK J.S., "Study of nonpolar m-plane InGaN/GaN multiquantum well light emitting diodes grown by homoepitaxial metal-organic chemical vapour deposition", *Applied Physics Letters,* no.91, p. 181120, 2007.

[LEF 01] LEFEBVRE P., MOREL A., GALLART M., TALIERCIO T., ALLÈGRE J., GIL B., MATHIEU H., DAMILANO B., GRANDJEAN N., MASSIES J., "High internal electric field in a graded-width InGaN/GaN quantum well: accurate determination by time-resolved photoluminescence spectroscopy", *Applied Physics Letters,* no.78, p. 1252, 2001.

[LER 98] LEROUX M., GRANDJEAN N., LAÜGT M., MASSIES J., GIL B., LEFEBVRE P., BIGENWALD P., "Quantum confined Stark effect due to built-in internal polarization fields in (Al,Ga)N/GaN quantum wells", *Physical Review B,* no.58, R13371- R13373, 1998.

[NAK 94] NAKAMURA S., MUKAI T., SENOH M., "Candela-class high-brightness InGaN/AlGaN double-heterostructure blue-light-emitting diodes", *Applied Physics Letters,* no.64, p. 1687, 1994.

[NAK 96] NAKAMURA S., SENOH M., NAGAHAMA S., IWASA N., YAMADA T., MATSUSHITA T., KIYOKU H., SUGIMOTO Y., "InGaN-Based multi-quantum-well-structure laser diodes", *Japanese Journal of Applied Physics,* no.35, L74, 1996.

[NAK 98] NAKAMURA S., "The roles of structural imperfections in InGaN-based blue light-emitting diodes and laser diodes", *Science,* no.281, p. 956, 1998.

[NAK 99] NAKAMURA S., "InGaN-based violet laser diodes", *Semiconductor Science and Technology*, no.14, R27-R40, 1999.

[NAK 00] NAKAMURA S., PEARTON S., FASOL G., *The Blue Laser Diode. The Complete Story*, Springer, Heidelberg, 2000.

[NAR 97] NARUKAWA Y., KAWAKAMI Y., FUNATO M., SHIZUO FUJITA, SHIGEO FUJITA, NAKAMURA S., "Role of self-formed InGaN quantum dots for exciton localization in the purple laser diode emitting at 420 nm", *Applied Physics Letters,* no.70, p. 981-983, 1997.

[ODO 99] O'DONNELL K. P., MARTIN R.W., MIDDLETON P.G., "Origin of luminescence from InGaN diodes", *Physical Review Letters*, no.82, p. 237-240, 1999.

[SME 03] SMEETON T.M., KAPPERS M.J., BARNARD J.S., VICKERS M.E., HUMPHREYS C.J., "Electron-beam-induced strain within InGaN quantum wells: false indium "cluster" detection in the transmission electron microscope", *Applied Physics Letters,* no. 83, p. 5419-5421, 2003.

[SON 06] SONDEREGGER S., FELTIN E., MERANO M., CROTTINI A., CARLIN J.-F., SACHOT R., DEVEAUD B., GRANDJEAN N., GANIÈRE J.D., "High spatial resolution picosecond cathodoluminescence of InGaN quantum wells", *Applied Physics Letters,* no. 89, p. 232109-232111, 2006.

[TAK 00] TAKEUCHI T., AMANO H., AKASAKI I., "Theoretical study of orientation dependence of piezoelectric effects in Wurtzite strained GaInN/GaN heterostructures and quantum wells", *Japanese Journal of Applied Physics*, Part 1, no.39, p. 413, 2000.

[WAL 00] WALTEREIT P., BRANDT O., TRAMPERT A., GRAHN H.T., MENNIGER J., RAMSTEINER M., REICHE M., PLOOG H., "Nitride semiconductors free of electrostatic fields for efficient white light-emitting diodes", *Nature,* vol. 406, p. 865-868, 2000.

Chapter 4

Diode Processing

4.1. Introduction

Since material growth was developed, academic laboratories and particularly industrialists have focused on the development of the sequence of the technological steps in order to create a LED. Some of these technological steps remain classical (etching or contact deposition) but others are more specific to GaN LEDs, due to the properties of this material, like substrate removal by the laser lift-off technique. Historically, the processing of commercially available high power LEDs was optimized at each stage of the component (diode, phosphors, optics, packaging, protection against ESD). The current evolution tends toward a global reflection at the scale of the component and defines the direction of the p-type contact development.

In this chapter, different LED structures will be described: vertical diodes and lateral structure, with or without substrate removal. Before the study of the classical technological steps of the process, the evolution of the LED design will be detailed to show the advantages and drawbacks. Finally, the main characteristic of high power LEDs is the external quantum efficiency. Like the transfer function in electronics, an efficiency representing the output power-input power ratio can be defined at each stage of the component.

Chapter written by Philippe GILET.

The aim of the chapter is to describe the fabrication process of efficient monochromatic LEDs. The issues concerning the phosphors will be covered in Chapter 5.

Historically, each sub-efficiency of the component has been optimized separately from the others. This has boiled down to optimizing the sum of all the sub-efficiencies. Now, the next step is to consider the entire component and not to study the sub-efficiencies as independent variables, but to optimize their product.

Considering the scale of the commercial issues, as soon as the p-doping was resolved (see Chapter 3), research in solid state lighting grew exponentially not only in academic laboratories but also in the R&D centers of the main lighting industries like OSRAM in Europe, Philips-Lumiled and CREE in the USA, and Nichia or Toyodai Gosei in Asia. In order to better protect their development, these firms have a very aggressive intellectual property (IP) policy, which leads to frequent trials. Thus, to develop its lighting technology, each firm has to take into account not only the technical criteria but also all the patents of its competitors or buy a very expensive trading license. Also, each new commercial solution arises from a compromise between the improvements of the LED performances, the costs and sometimes the restrictions due to IP. However, thanks to the effort in terms of means and manpower of these firms, significant breakthroughs in solid state lighting technology mainly come from the industrial R&D departments, even if it is very difficult to know the current progress of their research precisely. Finally, it should be noted that the following results stem from publications in which each team presents its results in the best way. The difficulty in knowing the real issues remains, especially for reliability, as the announced results are not easy to check. This aspect will not be covered in this chapter.

4.2. Orders of magnitude

Currently, the chip size of LEDs is almost normalized: 350x350 μm^2 and 1x1 mm^2 for nominal currents of 350 mA and 1A, respectively. Larger chips begin to appear. Commercially available high power LEDs contain several chips (usually 4 or 6 chips) assembled during the packaging step of the process flow. In the best cases, the operation voltage is around 3 V (the GaN bandgap is 2.7 eV and 2.7 V is the theoretical lowest operation voltage). To achieve an external efficiency of 50%, a thermal power of about 0.5 W and 1.5 W has to be dissipated for 350 x 350 μm^2 and 1 x 1 mm^2 chips, respectively. This equates to a few hundred watts per square centimeter. The thermal management of the component is clearly a key point as important heating involves lifetime decreasing and the internal quantum efficiency drop of the diode. As a comparison, in current microprocessors, 80 to 100 W have to be dissipated for a chips area of 1.2x1.2 cm^2. This represents a power density of 60 W per square centimeter. Despite the use of a large heat sink (typically 25 cm^2), the device operates around 100°C! Thus, throughout the design of the electroluminescent device, the technological choices must be guided by thermal management, from the electrical injection within the junction (series resistance, bandgap engineering) to the packaging (heat sink). By way of an example, we have tested in our laboratory a commercially available high power LED chip. This 1x1 mm^2 unpackaged chip was placed on a glass plate and after a few minutes of operation under 700 mA, the chip was welded to the glass plate. This shows that the chip temperature reached several hundred degrees Celsius. However, although the efficiency decreased, the LED was still running! In addition, Figure 4.1 shows the degradation of the electric contact during a continuous wave high current operation of a 1x1 mm^2 LED, one of the first fabricated at the CEA-LETI Minatec.

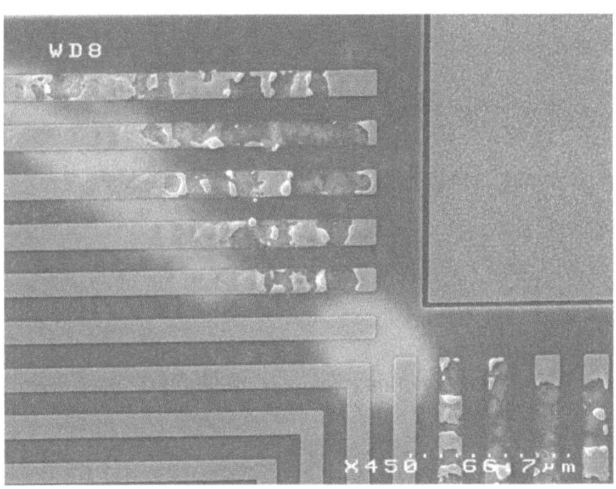

Figure 4.1. *Degradation of an ohmic p-type contact during a high current injection (1A in 1mm^2)*

For the GaN diodes, a threshold voltage of 2.7 V is imposed by the material bandgap. In its last progress report presented to the Department of Energy, the firm CREE [IBB 07] announced a threshold voltage of 3.15 V for a 700 mA current, which corresponds to a series resistance of 0.64 ohm. Achieving this value arises from the joint optimization of the injection design and the contact ohmicity.

Before detailing the diode configurations, we will give some orders of magnitude (Figure 4.2) regarding the external quantum efficiency and available optical output power for two emitting wavelengths [KRA 07].

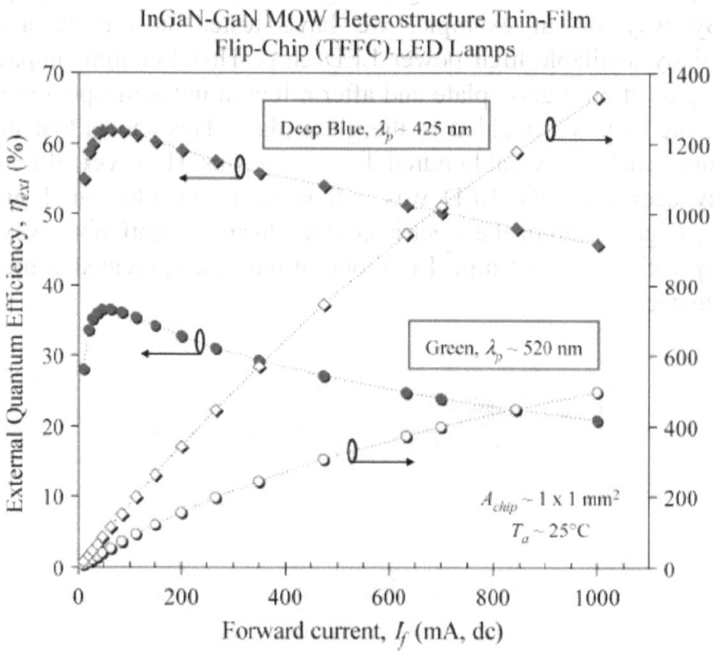

Figure 4.2. *External quantum efficiency and light output for emission at 425 and 520 nm. Chip size: 1x1mm² [KRA 07]*

Figure 4.2 presents the external quantum efficiency (EQE) and the light output, measured for two diodes in continuous wave operation and without active thermal management. The configuration of these diodes is given below. For the diode emitting in the blue range, the EQE is around 56% at 350 mA and achieves 62% at low injection. This component, which delivers a light output around 1.3 W at 1 A (continuous wave), is called "Watt class". For the diode emitting in the green range, the EQE is only 29% at 350 mA and reaches 36% at low injection. The maximum light output is only 500 mW.

This example from the state of the art in June 2007 partially illustrates the issues that challenge scientists: on the one hand, to prevent a drop in efficiency while the current increases and, on the other hand, to create green diodes with similar efficiency to blue diodes.

Regarding the second issue, the studies focus on epitaxy. The aim is to realize In-rich InGaN-based multi-quantum wells.

The EQE drop is mainly due to the internal quantum efficiency (IQE) drop (at the junction level), related to the rise of the injected current and Joule heating. We tackle this problem by developing specific technologies, which will be described in this chapter.

4.3. Diode configurations

[KRA 07] clearly provides the state of the art of LED configuration for nitrides (Ga, Al, In-N) and phosphides (Al, Ga, In-P). Indeed, as shown in Figure 4.3, both types of material are necessary to cover the whole visible range.

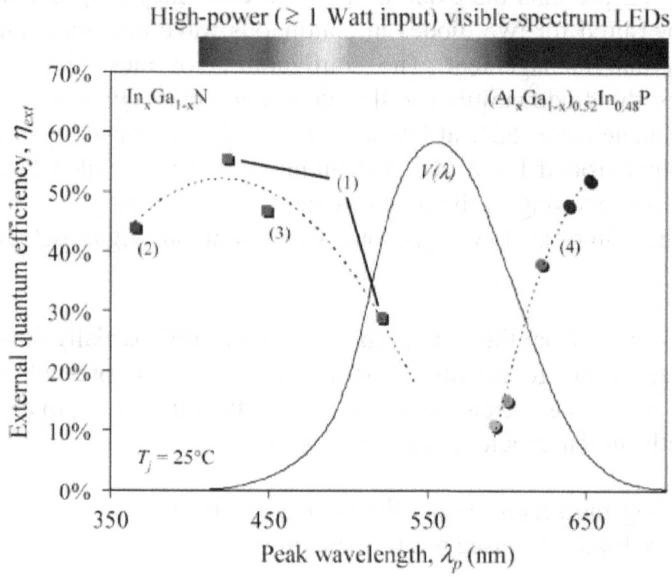

Figure 4.3. *[KRA 07] State-of-the-art external efficiency for high-power visible-spectrum LEDs at 25°C: (1) InGaN TFFC LEDs, 350 mA; (2) InGaN VTF LED, 1,000 mA; (3) InGaN CC LEDs employing patterned substrates; (4) production performance, AlGaInP TIP LEDs, Philips Lumileds Lighting Co., 350 mA. V (λ) is the luminous eye response curve from CIE. The dashed lines are visual guides*

This section only deals with the wide bandgap nitrides, which require different configurations compared to the phosphide-based system. Indeed, phosphides are grown on GaAs or GaP substrates, whereas nitrides are grown on sapphire which is electrically and thermally insulating.

The American firm CREE (www.cree.com) works with SiC substrates (good electrical and thermal conduction) for active layer growth. Historically, the configuration of the first nitride-based LEDs was like that of phosphide-based LEDs. For instance, the truncated inverted pyramid shape [EDM 04] makes it possible to exploit the transparency and the good electrical and optical conductivity of the substrate.

Later, in order to enhance the performances of the components and despite the interesting properties of SiC, the diode configurations are more and more similar to those developed for structures grown on sapphire.

Indeed, as explained below, in the current configurations, the substrate is removed to improve the device performance. Before describing the process flow in detail, the different types of diode configuration will be presented.

4.3.1. *Conventional chip (CC)*

In this "epi-up" configuration, the cathode is fabricated by locally etching the p-doped layer to deposit the n-contact on the thus-revealed n-doped layer. As the Mg-doped p-GaN has a very weak electrical conductivity (~ 1 Ω.cm), it is necessary to deposit a semi-transparent top contact on a large area in order to optimize the uniformity of the current injection.

The drawbacks of this structure are weak light extraction efficiency (absorption by the semitransparent electrode), non-uniform current injection which leads to current-crowding, Joule heating (thermal resistance) and finally a shorter lifetime.

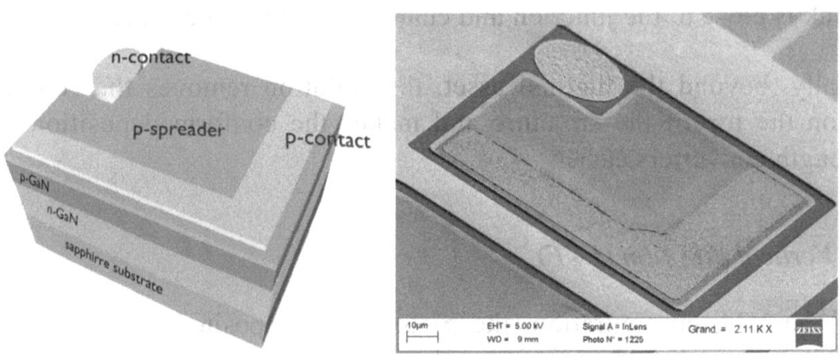

Figure 4.4. *Diagram and scanning electronic microscope image of a LED chip in CC configuration, fabricated at CEA-LETI*

4.3.2. *Flip chip (FC)*

The issue of CC configuration lies in the compromise between a good light extraction and a uniform current injection. Parallel to the increase of the injected current, enlarging the thickness of the semi-transparent contact, which is necessary to prevent a too high heating, reduces the quantity of extracted light by absorption. In the FC configuration, a reflective metal layer ensures a good p-contact on a large area of the diode and enhances the light extraction. In this case, the light is emitted through the substrate.

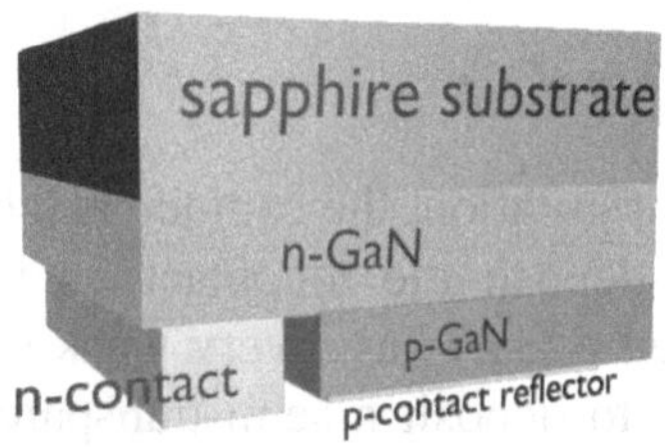

Figure 4.5. *Diagram of the FC configuration*

In addition, the structure is bonded on a low thermal resistance material, like silicon or ceramic. Thus, in this configuration, the thermal heat-sink is close to the junction and enables a good heat extraction.

Finally, beyond the thermal asset, this solution removes the contact wires on the top of the structure and makes the conform deposition of wavelength converters easier.

4.3.3. *Vertical thin film (VTF)*

For the VTF configuration, the p-contact is deposited on the whole diode area. Then the structure is bonded as in the previous configuration on a host substrate. The next step consists of removing the substrate to reveal the n-type layer for n-contact deposition. The substrate is removed by the Laser Lift-Off (LLO) technique, which will be described below.

In this case, the n-type contact is deposited in the center of the top surface of the device and the electrical resistivity of the n-type GaN (GaN: Si ~ 0.005 Ω.cm) is low enough to enable the current spreading on the whole diode surface.

In addition, the surface roughening of the sample is possible after the LLO step and thus the light extraction coefficient can be increased.

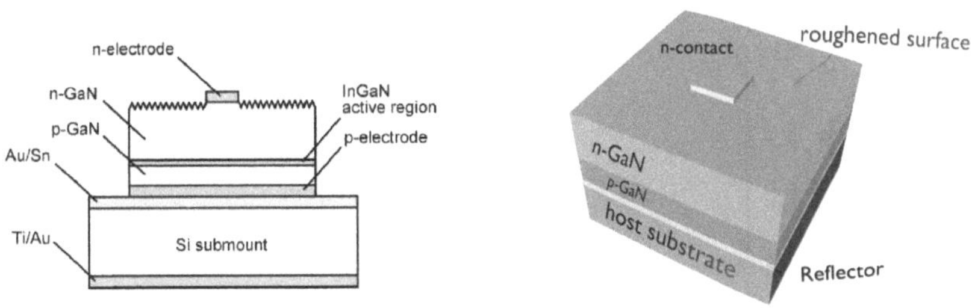

Figure 4.6. *Cross-section representation from [FUJ 04] and diagram of a VTF LED*

This solution leads to a good heat extraction, but the deposition of wavelength converters is difficult and the bonding wire on the top contact makes a shadow.

4.3.4. *Thin film flip chip (TFFC)*

This configuration corresponds to a combination of VTF and FC configurations to add the assets concerning thermal management, reliability of the technology (wireless) and easy deposition of phosphors. This solution consists first of processing the wafer in FC configuration and then removing the substrate using the LLO technique. This is a natural evolution of the FC configuration. Currently, the best LED performances are achieved with the TFFC configuration (see Figure 4.7).

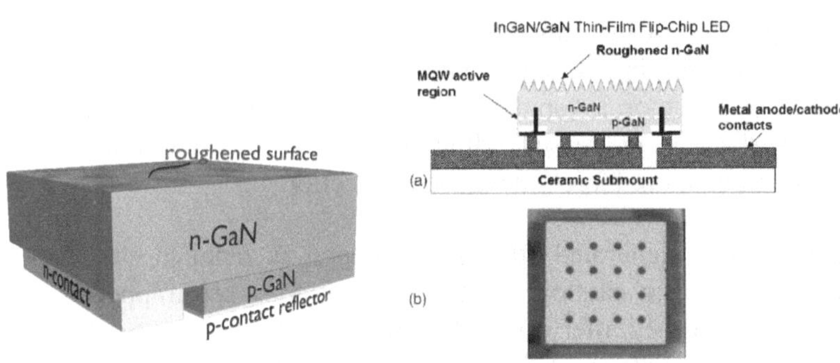

Figure 4.7. *Diagram of the TFFC configuration and representation of the device studied in [SHC 06]: (a) cross-sectional diagram and (b) photomicrograph*

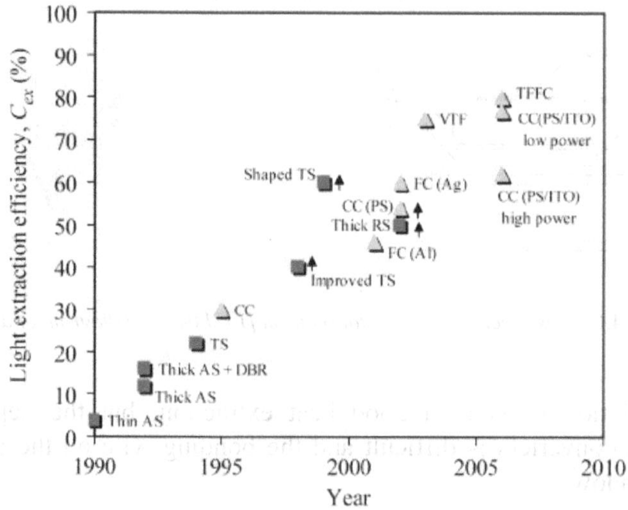

Figure 4.8. *[KRA 07] Evolution of extraction efficiency for AlGaInP (squares) and InGaN-GaN (triangle) LEDs. Upwards arrows indicate minimum estimates based on measures $\eta_{ext.}$ Other points are calculations. "FC(Al)" and "FC(Ag)" refer to the case of patterned sapphire substrate. "CC(PS/ITO)" refers to LEDs employing indium-tin-oxide contacts and a patterned sapphire substrate*

4.4. Light extraction at wafer level

Two main aspects are studied to enhance light extraction: first, the nature and the geometry of the electrodes, then texturing a part of the structure.

The deposition of thick metallic electrodes on top of the device obscures a part of the emitting surface. The solution consists of using a semi-transparent contact or a "spider's web"-like electrode. Thus, the lack of emitting surface is decreased while a uniform injection remains. These solutions will be more largely described in section 4.5.

Once light is generated (this phenomenon is detailed in Chapter 3), without any specific caution most of the photons are trapped inside the material and absorbed afterwards. According to the famous Snell-Descartes law ($n_1.\sin(i_1)=n_2.\sin(i_2)$), it is easy to demonstrate that only light rays emitted into a light escape cone of 23° (the refractive index of GaN is around 2.5) can escape from the material through the surface. Thus, neglecting reflection on the back side and on the side-walls of the device and assuming that the light emission within the quantum wells is omnidirectional[1], only 4% of light can escape. Most light is trapped within the material.

Several more or less developed solutions have been set to tackle this problem. As the light emission is (in first approximation) omnidirectionnal, a large part of the light is oriented toward the back side and is totally lost. Thus, the idea is to place a reflective surface at the emitted wavelength (450 nm) in order to angle the light to the emitting top surface.

This solution is based on silver or aluminum reflective layers. This will be explained in section 4.5. In addition, for FC, VTF or TFFC configurations, the distance between the quantum wells and the reflective layer can be in the order of the emitted wavelength (a few hundred nanometers). The extraction efficiency can be increased or decreased by the optical cavity effect in the structure [SHE 03], according to the thickness between the active layer and the reflective layer. Taking into account the accuracy of this thickness value, this solution does not sound realistic for mass production.

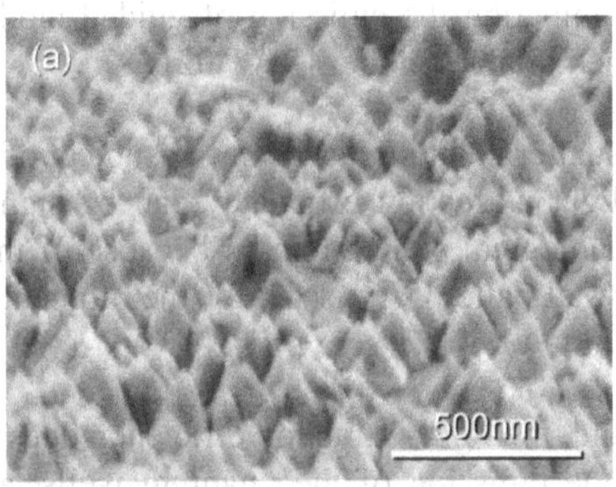

Figure 4.9. *Example of surface roughening from [FUJ 04]. SEM micrographs of an N-face GaN surface etched by a KOH-based PEC method*

Since 1973, Bergh [BER 73] showed that roughening the surface increases the light extraction efficiency (see Figures 4.6 and 4.7). This

1 To be precise, according to the hexagonal symmetry of GaN, the light emission is not exactly omnidirectional. The quantum well emission is slightly preferentially perpendicular to the growth plane. This contributes to the light extraction.

solution can be exploited whatever the configuration (CC, FC, VTF or TFFC), but would be more or less effective.

The principle is based on the randomness of the angle between the incident light ray and the emitting surface. During the numerous reflections, the probability of the light ray hitting the top emitting surface into the light extraction cone is increased thanks to the randomness of the surface. Then the light ray can escape to the free space.

At the end of the 1990s, several studies including photonic crystals were carried out. Thanks to a periodical modulation of the refractive index in the material, it is possible to create a photonic band structure, the formalism of which is similar to that of the electronic band structure in a crystal.

Thus, tracing the dispersion relation makes it possible to know the behavior of the optical modes of the structure and to know how to influence the light emission. To find out more about photonic crystals, see [LOU 03].

The ordered patterning of the surface as a photonic crystal can act in two ways: either by eliminating the emission of guided modes [FAN 97] or by coupling light to the radiation modes [RAT 02].

In practice, this latter solution has been the subject of more studies. With this method, the light extraction efficiency can be significantly increased. However, the technological strains remain because of the nanometric scale of the photonic crystal pattern, which is, for the moment, incompatible with mass production.

On the other hand, the light extraction efficiency does not cover the entire spectral range of the emission of the quantum wells, and this reduces its interest.

In addition, surface roughening can be placed inside the structure, between the substrate and the buffer layer. Then, it has two functions: first decreasing the number of dislocations in the active material by following the ELOG (Epitaxial Lateral OverGrowth, see Chapter 2) principle; secondly it creates a diffraction grating which suppresses the horizontal light guiding and angles the light outside the layer.

Figure 1.12. *Left: LED illumination of the Canon tower in Kowloon (Hong Kong) 14 lines with 30 LEDs per line. Each chip consumes 6 W and includes six junctions (2 red, 2 green and 2 blue), for a total power of 12.5 kWh per night [LED 07]; right: arches of the Pont Neuf bridge in Toulouse have been illuminated since November 2007 with a dynamic LED lighting (total power 1 kW)*

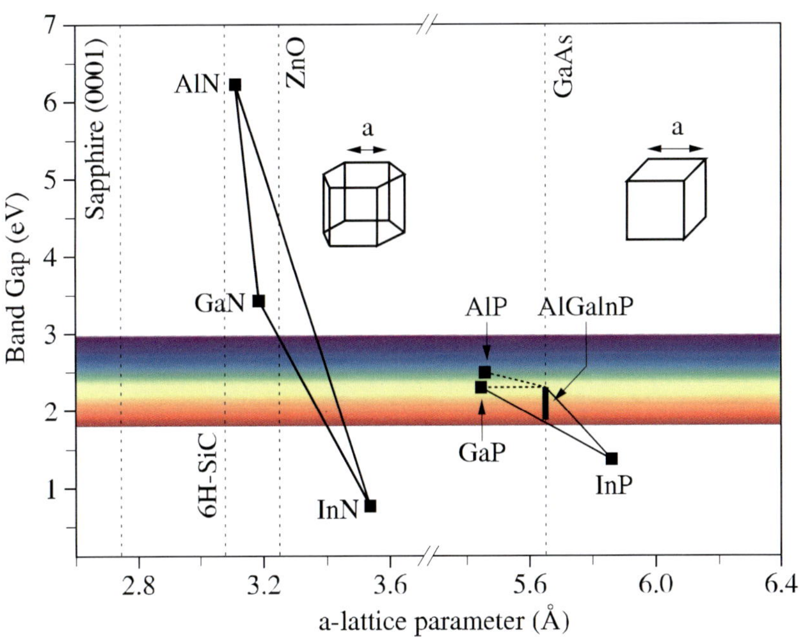

Figure 2.1. *Bandgap of nitrides (Al,Ga,In)N and phosphides (Al,Ga,In)P used for visible LEDs, as a function of their lattice parameter*

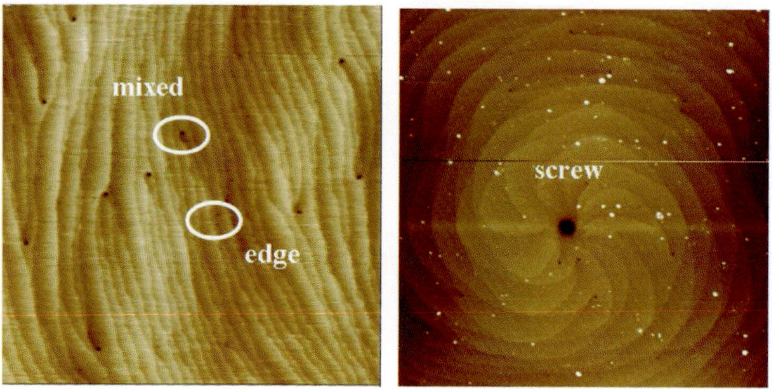

Figure 2.11. *AFM images (2x2 μm²) of MOVPE GaN over sapphire*

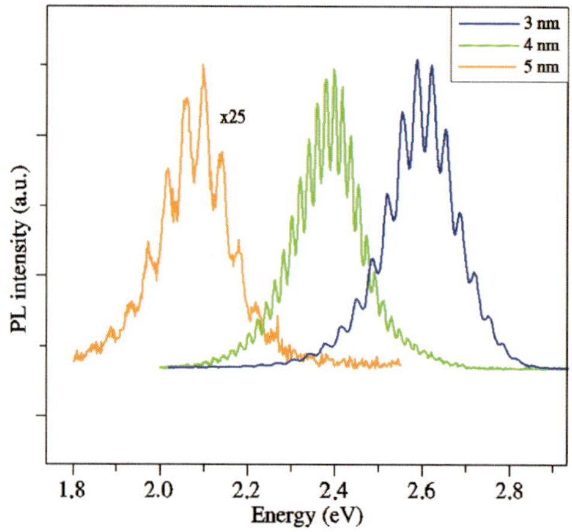

Figure 3.4. *PL spectra of In0.22Ga0.78N/GaN quantum wells emitting from blue to orange @ 300K*

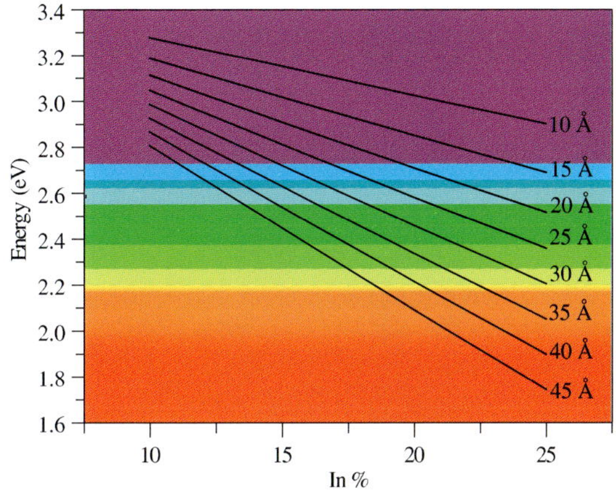

Figure 3.7. *InGaN/GaN quantum well transition energy considering both In% and well thickness*

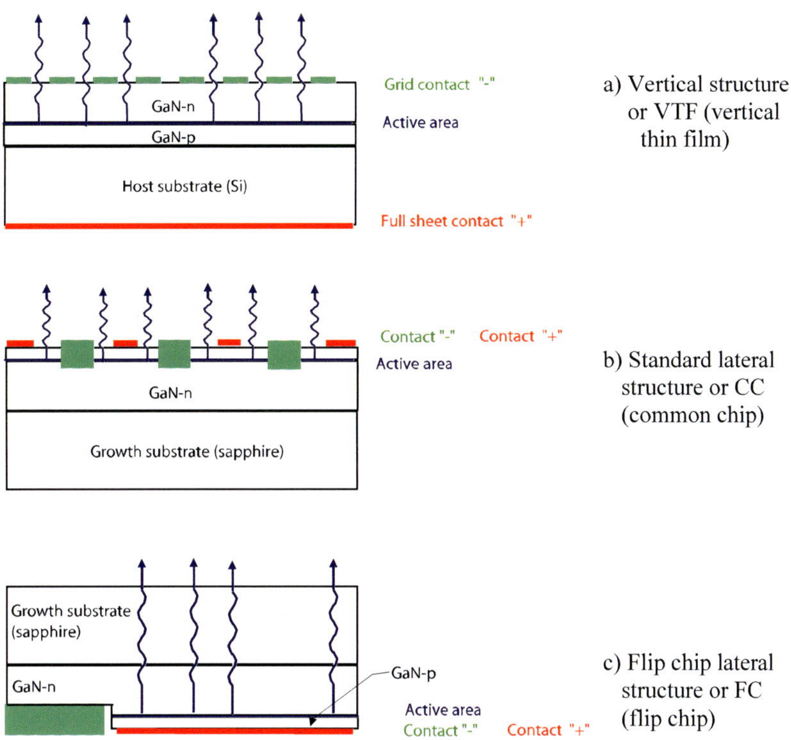

Figure 5.3. *Schematic representation of LED chips, vertical lateral and flip chip configurations*

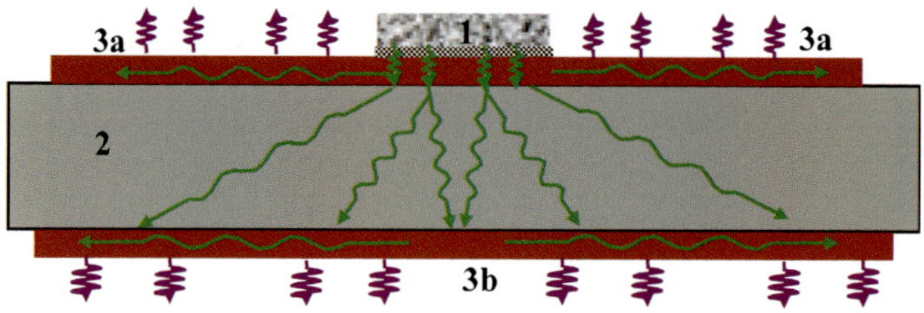

Figure 5.12. *Schematic representation of thermal spreading through conduction (green) and convection/radiation (purple)*

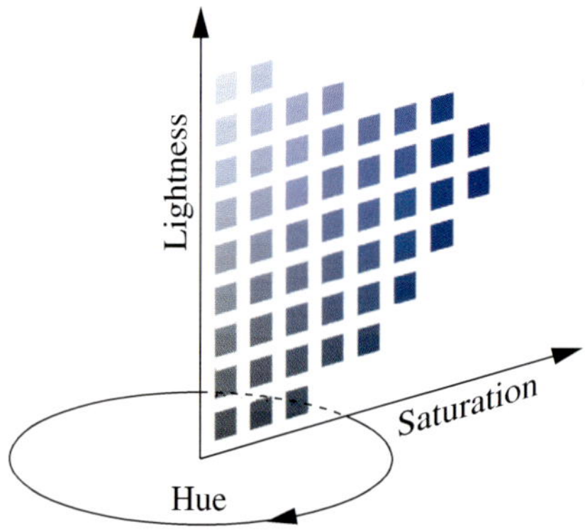

Figure 7.3. *Three dimension representation of the colors following the Munsel classification*

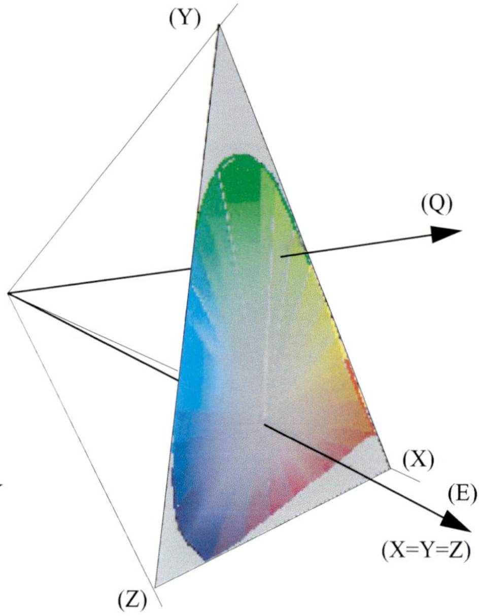

Figure 7.4. *Representation of colors in the color space XYZ. The equi-energy white point is at the center of the diagram*

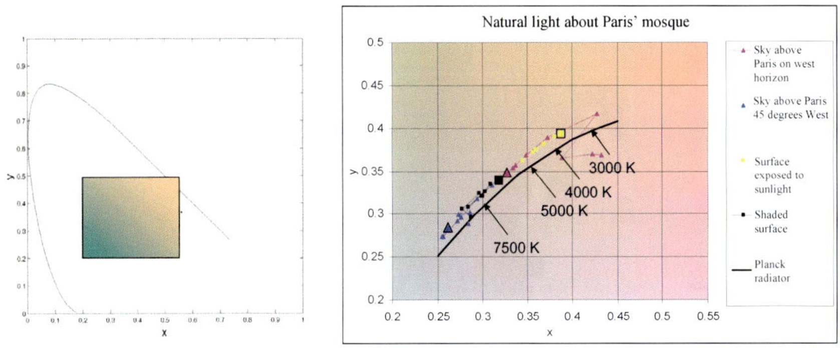

Figure 7.5. *Localization of the natural daylight on the chromaticity diagram **x, y** of the CIE*

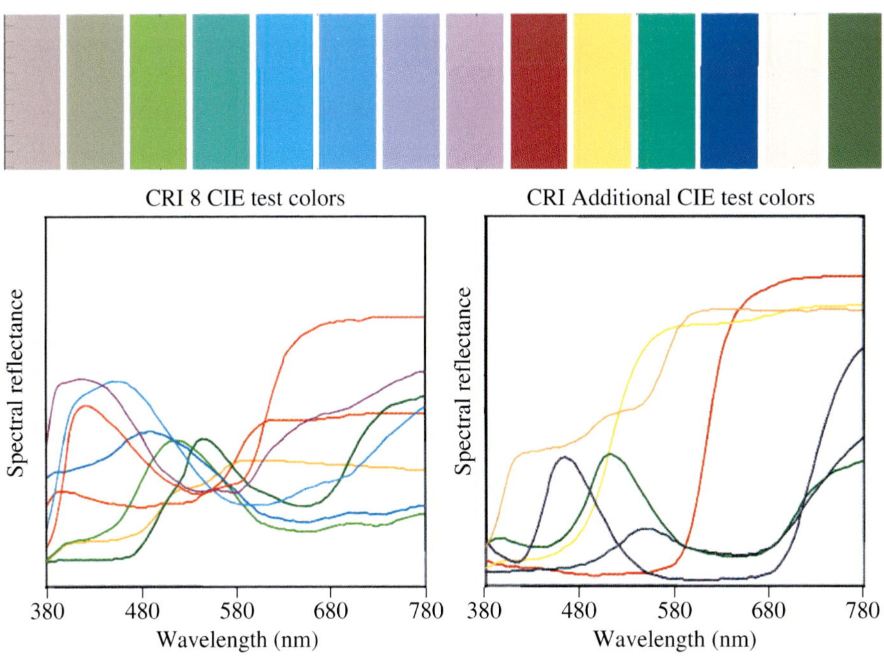

Figure 7.17. *Recommended test samples and reflectance factors for the CRI calculation*

Figure 7.19. *Example of color differences obtained with a tested source and the reference source, after the CRI calculation*

Figure 7.22. *Color discrimination test [VIE 07]*

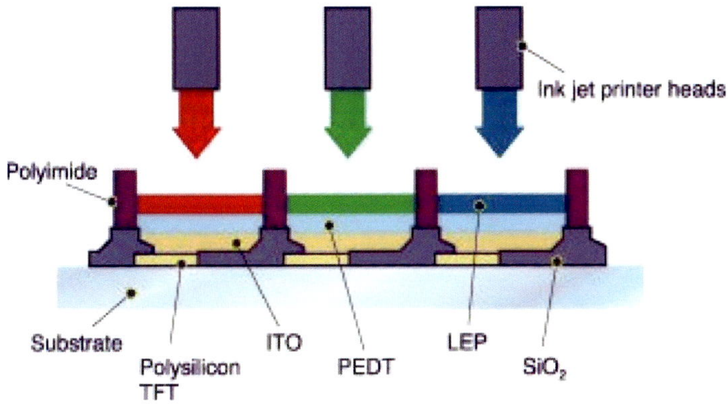

Figure 8.4. *OLED ink jet method for polymer deposition (here PEDT and LEP)*

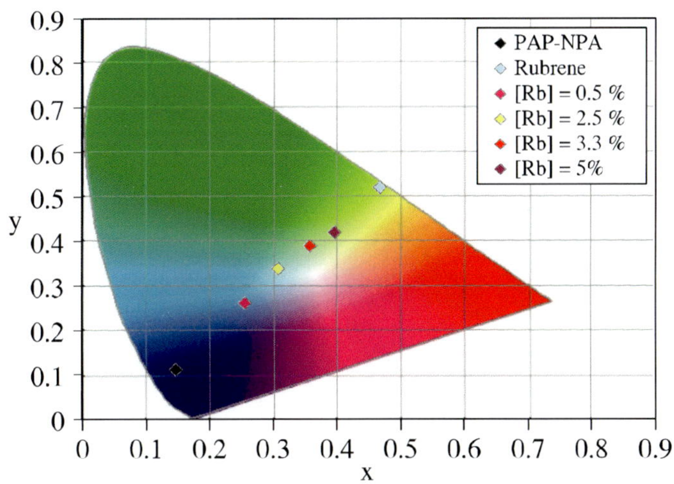

Figure 8.12. *Evolution of chromatic coordinates at fixed voltage according to the concentration of yellow emitting rubrene dopant for a PAP-NAP-based OLED*

Figure 8.17. *OLED 3" passive display on a PET plastic substrate (source: Pioneer 2002)*

Figure 8.19. *OLED-based lighting panel (source: General Electric 2005)*

To conclude, today's commercially available solutions combine the different solutions and are based on random surface roughening and reflective contacts. For TFFC [SHC 06] or VTF [IBB 03] configuration, the extraction efficiency is close to 80%.

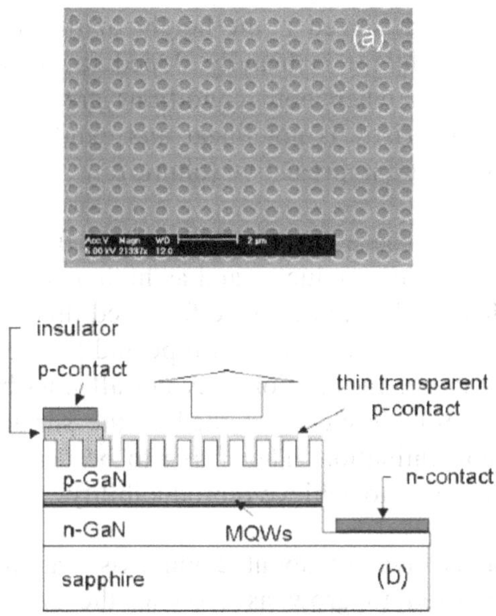

Figure 4.10. *SEM image of a photonic crystal at the surface of a LED and scheme of the structure from [KIM 05]*

4.5. Diode processing, etching, contact deposition

The different steps of diode processing will now be detailed. For more information, we recommend [PEA 99], despite its rather old publication date.

Whatever the chosen configuration, etching and n- and p-contact deposition are essential steps in the process flow. A good ohmic semiconductor-metal contact exhibits a symmetric, linear and temperature-independent I-V characteristic. In addition, the voltage drop at the electrode level has to be negligible compared to the voltage drop in the active area. Ohmicity is usually evaluated through the specific contact resistance in

Ωcm^2. Then, for reasons of reliability, the contact pads have to have a good adhesion on the material and a good mechanical stability.

It is difficult to obtain ohmic contact on wide gap materials and especially on p-type GaN because of its low doping level. In general, a good ohmic contact is achieved if the height of the Schottky barrier is low or if the semiconductor is heavily doped to enable the tunneling effect. The best case is when both conditions are met. Otherwise, the current-tension characteristic is not linear and exhibits a bend: then the contacts are rectifying and not resistive (ohmic).

The first rule is to use a metal whose work function is as low as possible for an n-type semiconductor and as high as possible for a p-type semiconductor. Most of the studies have followed this rule and many pad geometries have been realized on both n-type and p-type GaN, with more difficulties for the latter. In addition, beyond all possible metallization, the good contact quality is closely linked to the surface quality of the material and the contamination. They have to be controlled to improve the contact properties and to achieve reproducibility.

A thin amorphous layer of about 2 nm exists at the surface of the material. It is supposed to grow as soon as the epitaxial structure is exposed to air and it is mainly composed of gallium oxide (Ga_2O_3) and adsorbed carbon and hydrocarbon. This insulating layer involves an increase of 0.2 to 0.3 eV of the barrier height at the interface. Therefore, it is necessary to develop a treatment to remove the maximum number of impurities.

The simplest way is to make a chemical treatment on the samples. It is difficult to present the various studies exhaustively, but we will attempt to outlines the main tendencies, keeping in mind that the fabrication of good ohmic contacts is precious and guarded knowledge so that it is hard to obtain information about it. The most usual treatment is a dilute HCl solution which reduces the oxygen and carbon contamination on the surface. The analyses of similar treatments using *aqua regia* ($HCl:HNO_3$) at ambient temperature or boiling, BOE (*Buffered Oxide Etchant*, $HF:NH_4F$), dilute HF solution, $(NH_4)_2S_x$, or different sequences of dips have only shown limited results on the cleaning and/or the modification of the GaN surface properties. In most studies, the mechanism leading to the ohmicity is not well understood. For more information regarding surface treatments before metallization, see [HAR 01].

4.5.1. *N-type contacts*

As seen earlier, ohmic contacts on n-type GaN cannot be achieved with low work function (Φm) metal layers. Thus, titanium (Φm = 4.33 eV) or aluminum (Φm = 4.28 eV) and Ti/Al bi-layers are traditionally used. These metals have also other advantages: titanium's chemical properties give a good adhesion and aluminum is a good electrical conductor. However, they are both prone to oxidation. Usually they are covered by a gold layer to avoid this. Moreover, it is necessary to deposit a barrier to the diffusion of gold. Then, the contact layers are typically Ti/Al as adhesion and contact layers and Ti/Au, Pt/Au or Ni/Au as protection layers [CHE 04]. Beyond the metal layer sequence, thermal annealing has been the subject of many studies. The thermal treatment leads to an ohmic contact and also reduces the contact resistance. The optimum annealing temperature obviously depends on the contact composition and varies between 400°C and 900°C, while the duration varies from two to ten minutes [DUB 04]. For these temperature and time ranges, RTA (Rapid Thermal Annealing) enables flexibility and a good control of the process.

The best results have been reached with Ti/Al/Ni/Au – 115/220/40/50 nm after a thermal annealing at 600°C for two minutes: the specific contact resistance is $\rho c = 2.9 \times 10^{-5}\ \Omega.cm^2$ and even $10^{-6}\ \Omega.cm^2$.

4.5.2. *P-type contacts*

P-type contact on GaN is a complex challenge. Indeed, extrinsic p-doping is obtained by magnesium incorporation during growth, followed by high temperature activation.

Considering the activation energy of the magnesium dopant (around 200 meV), only about 1% of the Mg atoms are electrically active. As the maximum concentration of impurities is limited in order not to degrade the crystal, the maximum p-type doping we can hope for at ambient temperature is about 10^{17} cm^{-3}. This value is too low to obtain good ohmic contacts and injection uniformity.

In addition, the thermal annealing for contact ohmicity influences the photoluminescence intensity and the doping of the material [NAK 00]. In fact, according to the temperature and the atmosphere of the annealing, the photoluminescence signal can completely disappear and/or the doping level can significantly decrease. Therefore, the optimization lies in good ohmic contacts (which signifies a high doping level) concurrently to good photoluminescence efficiency.

Ideally, the metal used for the contact on p-type GaN has to have a high work function, about 6.5-7.5 eV. However, high work function metals are Pt, Ni, Pd and Au with work function values of 5.65, 5.15, 5.12 and 5.1 eV, respectively. Thus, it seems to be impossible to realize a real ohmic contact based only on the thermionic effect; the tunneling effect has to be promoted [ZEL 03]. Therefore, a specific effort has to be made regarding surface treatment and thermal annealing, but the first thing is to obtain good quality samples, the doping level of which is well controlled and high enough to achieve ohmic p-type contacts.

Most p-type contacts are stacks of some of the previously quoted metals. Usually, Ni/Au is used, but Ni/Pt/Au, Ti/Pt/Au, or Pd/Au are also used. Ni/Au – 20/100 nm stacking is exhaustively studied as it has good electrical properties and leads to specific contact resistance value of a few 10^{-6} $\Omega.cm^2$ [CHI 01].

The characteristic of the contact stacking is closely linked to its thermal annealing. Annealing has been tested in different atmospheres but only the oxygen atmosphere leads to ohmic p-contacts. The annealing has two effects: it enables the diffusion of Ni atoms in the upper GaN layer and also their oxidation [HOR 01]. The oxidation is essential to obtain ohmicity as the formed NiO behaves as a wide-bandgap p-type semi-transparent semiconductor [BOG 06]. Usually, these p-contact annealing are realized in air or in oxygen atmosphere between 500°C and 600°C.

It should be noted that annealing of p-contacts on GaN in O_2 atmosphere is not intuitive. Indeed, the diffusion of oxygen atoms (deep n-type dopant) could change the doping type at the surface or even "kill the luminescence" if the oxygen reaches the quantum wells. The optimization of contact annealing is known to be a delicate step in the process. Now, what about the geometry and the nature of the p-contacts for the different quoted configurations?

As seen previously, the CC LEDs (the simplest configuration) emit through the top face; this means through the p-contacts. Therefore, it is essential to find a trade-off between the homogenity of the electrical injection and the transparency of the surface, namely between the opaque metallic area and the transparent area. Two ways are then considered: either a limited area of thick opaque metallization (spider-web-like contact) or a thin semi-transparent contact over the whole area. One of the reasons of the large study of Ni/Au contact is the wide gap of NiO, which confers to it the transparency in the visible range. Other types of contact can also be used like Ru/Ni, Pd/Ni, but the Ni/Au-based process has been the easiest to develop. Thus, many devices include Ni/Au p-contacts. The transparency of this stacking is around 60-75% for a specific contact resistance of $5x10^{-4}$ $\Omega.cm^2$ [LIN 03].

Nevertheless, in order to further improve the performances, conductive oxides like ITO (indium tin oxide) [LIN 03] or ZnO [LIM 04]-[SON 03] make it possible to obtain a good conductivity and a higher optical transmission. These materials are called TCOs (Transparent Conducting Layers). Their optical transmission is higher than or equal to 80% for resistivity around 10^{-4} $\Omega.cm$ [KIM 04]. With these materials, the issue still lies in the realization of a good p-contact. The direct deposition of an ITO or ZnO layer does not lead to an ohmic contact, but a thin Ni layer between GaN and TCO significantly enhances the ohmic characteristic of the contact [HOR 01]. Specific contact resistances of $8.6x10^{-4}$ $\Omega.cm^2$ and 10^{-5} – 10^{-6} $\Omega.cm^2$ have been achieved for Ni/ITO and Ni/ZnO, respectively.

For LEDs in advanced configuration, the p-contact is buried and also has a light reflector function. In this case, classical Ni/Au contact is no longer suitable. Indeed, for the wavelengths of interest (400-450 nm), gold heavily absorbs the light. It is essential to use other metals like silver or aluminum. Silver-based metallic layers have a reflectivity up to 95% from 340 nm and along the visible range and the reflectivity of aluminum-based metallic layers reaches 90% in the whole wavelength range of the application. Even if silver has often been used, its electrical characteristics are not reliable [SON 05]. In addition, silver layers do not efficiently bind to GaN and their morphology deteriorates during the thermal annealing. Finally, ohmicity is not achieved and reflectivity is too low. A solution is to insert a thin Ni layer between GaN and silver [JAN 04] to obtain the ohmicity and to keep a high reflectivity. Some groups have tested Ni/ITO/Ag. Ni-ITO gives the transparent ohmic contact (see above) and the Ag layer plays the part of the reflector. In these cases, the ohmic contact stems from a trade-off between a low specific contact resistance value (around 10^{-5} $\Omega.cm^2$) and a high total reflectivity of the stacking.

4.6. Etching

In normal conditions, wide-bandgap nitride exhibits a high chemical inertia in chemical reactions.

Only boiling acid or basic solutions like H_3PO_4, H_2SO_4 or NaOH, KOH [WEY 00] lead to GaN etching. However, etching preferably appears on material defects like dislocations and gives a rough etched surface. Even if UV-assisted etching modifies the profiles, it is not efficient for mesa or trench processing.

Nevertheless, this strong etching anisotropy is used for the last step of the process. The surface of the device is roughened to improve the light extraction, especially in VTF and TFFC configurations.

Dry etching is essential for the fabrication of GaN-based devices. Dry etching development has been focused on high etching rate for the realization of trenches or mesas with anisotropic profiles and a smooth bottom surface, and independently of the GaN alloy composition (GaN, InGaN, AlGaN).

GaN dry etching remains a difficult step because of the high chemical inertia of the material and the high bond energy of N-III components compared to the other composed semiconductors. For example, the bond energy of GaN is 8.92 eV per atom compared to 6.52 eV per atom for GaAs [PEA 99]. Since the first demonstration of feasibility [ADE 93], many studies were made with different types of etching chamber (RIE, ICP, ECR) and of gases, especially halogen-based gases (Cl_2, BCl_3, ICl, IBr, BI_3, BBr_3, etc.) combined with inert ions like Ar and N. The current processes lead to anisotropic etching, vertical sidewalls and etching rates of about 1 µm/min. These results are based on the chemical reaction of chlorine on the element III combined with the mechanical action of high energy ion-induced bombardment. To obtain high resolution patterns and smooth bottom surfaces, chemical and physical contributions of the reaction have to be well balanced. Each type of etching chamber requires a specific study. Parallel to the pattern, the impact of the etching process on the surface on which the n-type contact is deposited in CC and TFFC configurations has to be minimized. For a too violent etching process, the material can be damaged and sometimes becomes amorphous, and it is no longer possible to realize a good ohmic contact. As usual, a trade-off between the etching rate, the etching profiles and the material deterioration is necessary.

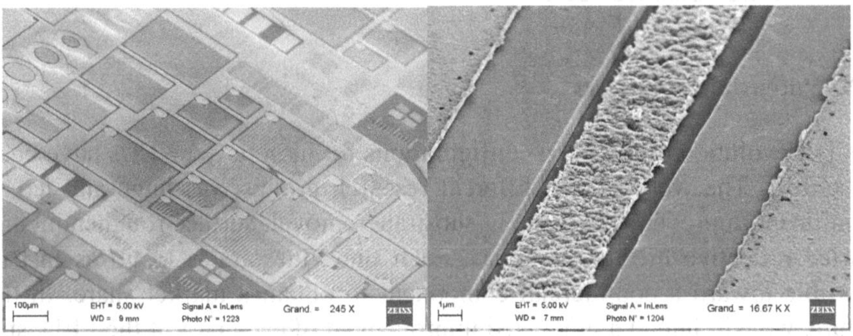

Figure 4.11. *Scanning electronic microscopy images from CEA Léti MINATEC. Left: view of a part of a wafer under process; right: detail of an n-contact in the middle and p-contact on top of a mesa on both sides*

4.7. Substrate removal

As previously described, the realization of the advanced configurations requires the removal of the sapphire or silicon carbide substrate. Although the technical solutions exist, this step remains a limitation for the low cost mass fabrication of high power LEDs. Sapphire is a very hard material which cannot be removed by polishing. The solution consists of lighting the back side of the sample with an excimer laser (KrF) at 248 nm. Sapphire is transparent at this wavelength, but GaN is not. With the absorption of the laser beam, a split zone, due to the formation of nitrogen bubbles and gallium droplets, appears in the buffer by the effect of demixion induced by the intense local heating. A post-insulation thermal annealing makes it possible to remove the active layer from the substrate. Although high power lasers with large spot size exist, this step is both delicate to control and time-consuming. Indeed, to reach a sufficient optical power density, the insulation is not made on the whole wafer but the laser with a rectangular spot scans the surface of the sample. Once the substrate is removed, the residual gallium droplets are eliminated by HCl etching. This step is often followed by photo-enhanced electrochemical etching (PEC) [MIN 96] to roughen the surface in order to increase light extraction (see section 4.4 and Figure 4.9). In the case of the SiC substrate (polytype 6H), this technical solution is not efficient because the SiC bandgap is 3 eV, which corresponds to 413 nm and the SiC is not transparent at the excimer laser emitting wavelength. The technological process mainly used by the American firm CREE today remains a well guarded secret.

4.8. Potential evolutions

The evolution of the main configurations for high power LEDs have been described. The work of the different research groups was huge and led to significant improvements, namely substrate removal, surface roughening and buried reflective metallic contacts. Each time, the aim is to go as fast as possible toward a commercial component to obtain a market share. In the future, we will intend a graduated improvement of each part of the LED, still by considering the effect of each change on the entire device performances. We naturally think about the contact geometry and the different materials to use. n-type contact seems to be well-controlled. On the other hand, p-type contact is perfectible. The main issue of these improvements concerns the thermal management, either to decrease heating (to lower the series resistance, etc.), to enhance the dissipation of the generated calories (materials with good thermal conductivity, etc.) or to improve the light extraction (to reduce the internal heating, etc.). Today it is clear that the substrate removal is essential. Studies still remain to be made to enhance the process flow, especially in terms of efficiency and fabrication cost. In order to improve the external quantum efficiency (see section 4.2), it is necessary to focus on the packaging and to be concerned about associating different materials bringing interesting physical properties, for instance a good reflectivity, a good thermal conductivity and close thermal dilation coefficients. Thus, the realization of components with high performances lies in the right combination of the epitaxial structure and the technological process flow at the chip level and for the packaging. A modification in one of these parts implies a modification in one or both of the other parts.

Of course, parallel to this mainstream, well-established and frozen in the OIDA road map, several more upstream studies are in progress to test and bring innovative solutions. We can quote the use of nanowires as upper contact surface to, first, improve light extraction and secondly achieve a good ohmic p-type contact on GaN. Indeed, these few hundred nanometer-sized new objects usually have a good crystalline quality and can exhibit interesting properties in terms of light guiding, like an optical fibee, and thus facilitate the light extraction. The studies in progress show a certain potential but the commercially available components are so mature that a real asset has to be proved to overturn the current technologies.

4.9. Bibliography

[ADE 93] ADESAIDA I., MAHAJAN A., ASIF KHAN M., OLSEN D.T., KUZNIA J., "Reactive ion etching of gallium nitride in silicon tetrachloride plasmas", *Applied Physics Letters,* no.63, p. 2777, 1993.

[BER 73] BERGH A.A., HILL M., SAUL R.H., PLAINS S., US Patent, no.3739217, 1973.

[BOG 06] BOGART K., CROFTON J., "Calculations and measurements of contact resistance of semi-transparent Ni/Pd to p-GaN", *Journal of Electronic Materials*, vol. 35, no.4, 2006.

[CHE 04] CHEN Z.Z., QIN Z.X., HU C.Y., HU X.D., YU T.J., TONG Y.Z., DING X.M., ZHAN G.Y., "Ohmic contact formation of Ti/Al/Ni/Au to n-GaN by two-step annealing method", *Materials Science and Engineering B,* no.111, p. 36-39, 2004.

[CHI 01] LEE C.-S., LIN Y.-J., LEE C.-T., "Investigation of oxidation mechanism for ohmic formation in Ni/Au contacts to p-type GaN layers", *Applied Physics Letters*, no.79, p. 3815, 2001.

[DUB 04] DUBOZ J-Y., "Matériaux semi-conducteurs à grand gap III-V à base de GaN", *Techniques de l'Ingénieur*, Electronics series, E 1 995-1, E 1 995-24.

[EDM 04] EDMOND J., ABARE A., BERGMAN M., BHARATHAN J., LEE BUNKER K., EMERSON D., HABERRERN K., IBBETSON J., LEUNG M., RUSSEL P., SLATER D., "High efficiency GaN-based LEDs and lasers on SiC", *Journal of Crystal Growth*, vol. 272, p. 242-250.

[FAN 97] FAN S., VILLENEUVE P., JOANNOPOULOS J., SHUBERT E., "High extraction efficiency of spontaneous emission from slabs of photonic crystals", *Physical Review Letters*, no.78, p. 3294, 1997.

[FUJ 04] FUJII T., GAO Y., SHARMA R., HU E.L., DENBAARS S.P., NAKAMURA S, "Increase in the light extraction efficiency of GaN-based light-emitting diodes via surface roughening", *APl*, vol. 84, no.6, p. 855-857, 2004.

[HAR 01] HARTLIEB, P. J., Electrical, chemical, and structural characterization of the interface formed between Ni/Au and Pd/Au ohmic contacts and cleaned p-type GaN(0001) surfaces, PhD thesis, North Carolina State University, Raleigh, NC, USA, 2002.

[HOR 01] HORNG R., WUU D., LIEN Y., LAN W., "Low resistance and high-tranparency Ni/indium tin oxide ohmic contacts to p-type GaN", *Applied Physics Letters*, no.79, p. 25-2927, 2001.

[IBB 03] IBBESTON J., "High efficiency LED lamp for solid state lighting", De-FC26-03nt41943 final report.

[JAN 04] JANG H.W., LEE J., "Mechanism for ohmic contact formation of Ni/Ag contacts on p-type GaN", *Applied Physics Letters*, no.85 , p. 5920-5922, 2004.

[KIM 04] KIM S.Y., LEE J.L., "Highly reflective and low-resistant Ni/Au/ITO/Ag Ohmic contact on p-type GaN", *Electrochemical and Solid State Letters*, vol. 7, no.5, G102-G104, 2004.

[KIM 05] KIM D.-H., CHO C.-O,. ROH Y.-G, JEON H., PARK Y. S., CHO J.,.IM J. S, SONE C., PARK Y., CHOI, W. J., PARK Q-H., "Enhanced light extraction from GaN-based light-emitting diodes with holographically generated two-dimensional photonic crystal patterns", *Applied Physics Letters*, no.87, p. 203508, 2005.

[KRA 07] KRAMES M.R., SHCHEKIN O., MUELLER-MACH R., MUELLER G. O., ZHOU L., HARBERS, G., CRAFORD G., "Status and future of high power light emitting diodes for solid state lighting", *Journal of light display Technology,* vol. 3, no. 2, p. 160-175, juin 2007.

[LEE 01] LEE C., LIN Y., "Investigation of oxidation mechanism for ohmic formation in Ni/Au contacts to p-type GaN layers", *Applied Physics Letters,* no. 79, p. 3815, 2001.

[LIN 03] LIN Y., CHANG S., SU Y., TSAI T., CHANG C., SHEI S., KUO C., CHEN S., "InGaN/GaN light emitting diodes with Ni/Au, Ni/ITO and ITO p-type contacts", *Solid State Electronics*, no. 47, p. 849-853, 2003.

[LIM 04] LIM J., HWANG D., KIM H., OH J., YANG J., NAVAMATHAN R., PARK S., "Low resistivity and transparent indium-Tin-Oxide-doped ZnO ohmic contact to p-type GaN", *Applied Physics Letters,* no.85, p. 6191-6193, 2004.

[LOU 03] LOURTIOZ J. M., BENISTY H., BERGER V., GERARD J. M., MAYSTRE D., TCHELNOKOV A., *Les cristaux photoniques ou la lumière en cage*, Hermes Science, 2003.

[MIN 96] MINSKY M.S., WHITE M., HU E.L., "Room-temperature photoenhanced wet etching of GaN", *Applied Physics Letters,* no.68, p. 1531, 1996.

[NAK 00] NAKAMURA S., PEARTON S., FASOL G., *The Blue Laser Diode: The Complete Story*, Springer, Berlin-Heidelberg, 2000.

[PEA 99] PEARTON S.J., ZOPLER Z.C., SHUL R.J., REN F., "GaN: processing, defect and devices", *Journal of Applied Physics,* vol. 86, no.1, p.1-78, 1999.

[RAT 02] RATTIER M., BENISTY H., STANLEY R., CARLIN J.F., HOUDRE R., OESTERLE U., SMITH C., WEISBUSH C., KRAUSS T., "Toward ultrahigh-efficiency aluminum oxide microcavity light-emitting diodes: guided mode extraction by photonic crystals", *IEEE Journal of Selected Topics of Quantum Electronics*, vol. 8, no.2, p. 238, 2002.

[SHC 06] SHCHEKIN O.B., EPLER J.E., TROTTIER T.A., MARGALITH T., STEIGERWALD D.A., HOLCOMB M.O., MARTIN P.S., KRAMES M.R., "High performance thin-film flip-chip InGaN-GaN light emitting diodes", *Applied Physics Letters*, no. 89, p. 071109, 2006.

[SHE 03] SHEN Y. C., WIERER J. J., KRAMES M. R., LUDOWISE M. J., MISRA M. S., AHMED F., Y. KIM A., MUELLER G. O., BHAT J.C., STOCKMAN S.A., MARTIN P.S., "Optical cavity effects in InGaN/GaN quantum-well-heterostructure flip-chip light-emitting diodes", *Applied Physics Letters,* no. 82, p. 2221-2223, 2003.

[SON 03] SONG J., KIM K., PARK S., SEONG T., "Highly low resistance and transparent Ni/ZnO ohmic contacts to p-type GaN", *Applied Physics Letters*, no. 83, p. 479-481, 2003.

[SON 05] SONG J.O., KWAK J.S., PARK Y., SEONG T.Y., "Ohmic degradation mechanisms of Ag contacts on p-type GaN", *Applied Physics Letters*, no.86, p. 062104, 2005.

[ZEL 03] ZELLWEGER C., Realization of GaN-based light emitting devices, Thesis, EPFL Lausanne, Switzerland, 2003.

Chapter 5

Packaging

5.1. Introduction

In electronics, packaging expresses the general way to connect the integrated circuit's microelectronics to the user's macro electronics world. Four packaging levels are usually identified and can be illustrated for lighting applications as follows.

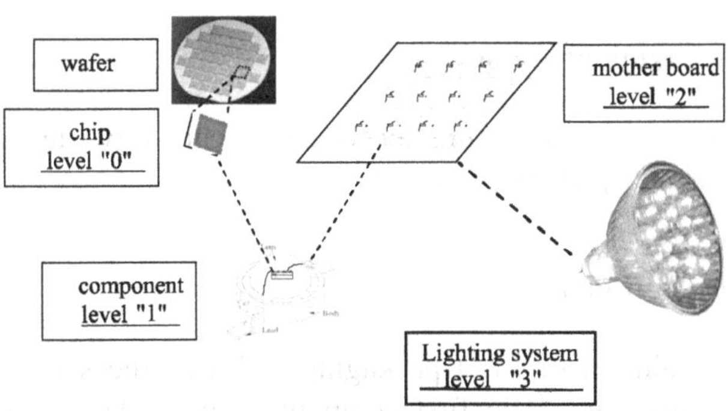

Figure 5.1. *Illustration of the 4 stages of packaging: level "0" = interconnections in the chip, level "1" = interconnections between the chip and its casing (component), level "2" = interconnections between several packaged components on a motherboard, level "3" = interconnections between one or several motherboards and integration into a lighting system [AMA 07]*

Chapter written by Adrien GASSE.

In the specific light emitting diode case, the wafer interconnections from level "0" are discussed in Chapter 4. In this chapter, we will mainly depict level "1", which allows us to develop a single component based on one or several light-emitting diode (LED) chips.

LED packaging should not only ensure an optical function but, beyond this main role, should also deal with other technical issues like: electrical interconnections, heat dissipation, mechanical strength, environment stability, reliability, etc. In addition to these functional and performance aspects, other features like cost, volume, manufacturing constraints and norms compliance have to be taken into account.

In the following section, we will present a quick historical review from the first manufacturing processes of LED components to today's methods describing the main LED packaging groups. We will show in section 5.3 problems related to thermal management. In fact, even though the light output efficiency of LEDs is significantly higher than incandescent or fluorescent bulbs, about 80% of the electric power input in the LED is still dissipated as heat. In section 5.4, we will discuss primary optics issues (encapsulant and lens) with light extractions and related materials. Finally, in section 5.5, we will summarize the different characteristics related to packaging given in the components' technical data sheet.

5.2. Different packaging processes

For more information concerning terms and technology standards in electronic packaging, see [SAI 05, HAR 04].

5.2.1. *Historical background*

The historical evolution of LED packaging followed the same path as other electronic components. The first components were developed to be bondable with each other, then connectable to a common motherboard using pins that go through it. Finally leadform components were attached on the surface of the motherboard (Surface Mount Technology (SMT)) to handle more components on it. This evolution is illustrated in Figure 5.2.

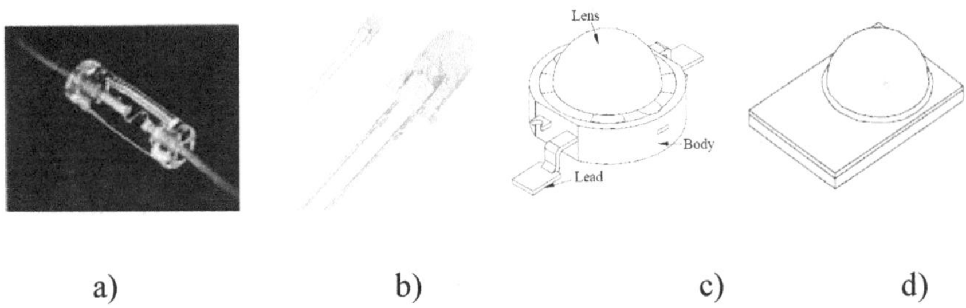

Figure 5.2. *Historical evolution of LEDs fabrication technologies: a) first LED component [HOL 62], b) LED with pins, c) SMT standard "leadform" LED, d) SMT "leadless" LED*

In 1962, Nick Holonyak Jr. and Sam Bevacqua from General Electric (GE) realized the first packaging for a red light emitting diode (a) [HOL 62]. During the 1970s, LEDs were packaged using two connection pins and were in a transparent or colored epoxy casing (b). In the 1990s and especially for Hig-Brightness LEDs, pin casings were dropped for SMT packages (c) with wide connections strips allowing upper surface mounting, a better heat dissipation, a higher current density, and lens type optics enhancement ("leadform" component). Since the end of the 1990s and in the current decade, two main trends have been observed: 1) highly integrated, small and low profile SMT components, e.g. for portable applications, and 2) high power components, not necessarily very small and thin but with a heat sink located in the casing, right below the chip, to improve thermal dissipation. To satisfy mass production and industrialization's needs for miniaturization, leadless components, without strips, appeared.

5.2.2. *From the wafer to the chip*

Nowadays, chips are produced on large diameter wafers, up to 100, even 200 mm. Electrical contact configuration and the emitted light orientation restrict the packaging of LEDs. The following illustration schematically represents vertical and lateral diode structures. To simplify, the main semiconductive element of the diode is called "GaN". Two types of LED chips have a lateral structure configuration, namely common chip and flip chip, where both contacts are on the same side. In the first configuration light is emitted through the contact's side, whereas it is emitted on the opposite side in the case of the flip chip configuration (see Figure 3.5).

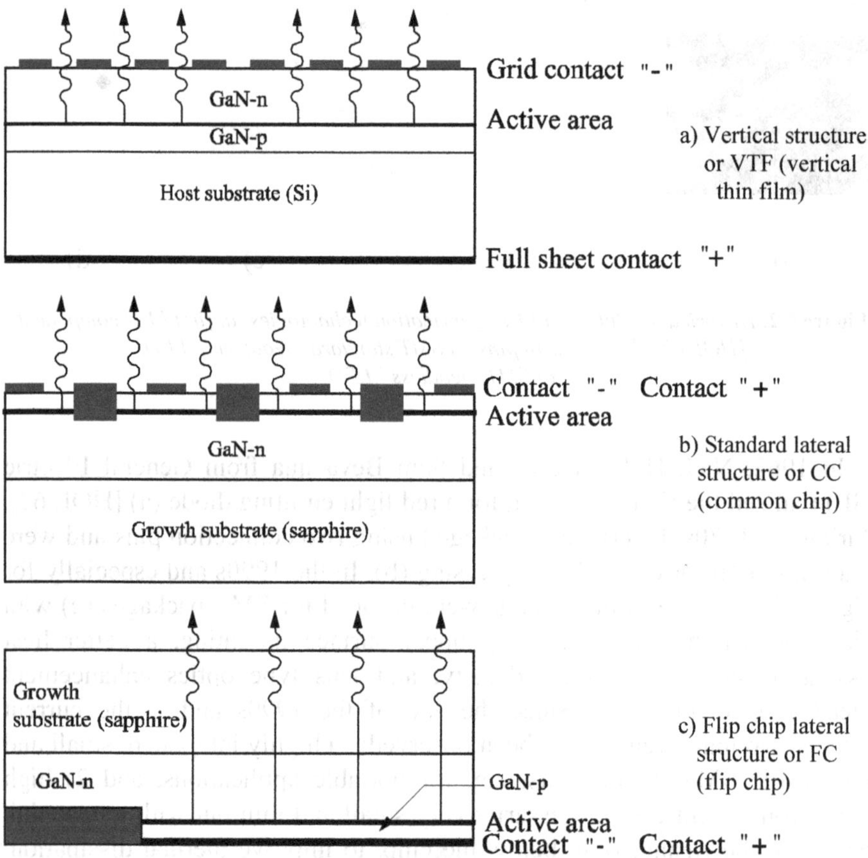

Figure 5.3. *Schematic representation of LED chips, vertical lateral and flip chip configurations (see color plate section)*

In the vertical LED case, the host substrate base is electrically conductive. It is generally doped silicon or doped germanium; the wafer's back side is usually coated with a thin metallic layer of a few hundred nanometers if the chip is glued, and occasionally, a thicker solder layer of a few microns if the chip is to be soldered. Some of the most standard metals used are bi-layers such as Ti/Ni, Ti/TiW, Ti/TiN, TiN/W, Ti/WN, the first layer being a bonding layer to the substrate and the second layer being a diffusion barrier against solder metallic alloys which could come and react with the LED's active layers. Most of the time, a third thin layer of protecting gold is applied to prevent oxidation of the metallic bond. The solder layer is usually 80/20 Au-Sn. This alloy has a melting point of 280°C, higher than most other typical leadfree alloys (Sn-Ag, Sn-Ag-Cu, etc.) or Sn-Pb used to solder components on motherboards. For more details on solder alloys, see [JAC 93]. In the lateral LED case, the substrate base is often the growth substrate, sapphire. The sapphire's back side is either metallized or not in the common chip configuration. However, the sapphire's backside is definitely not metallized in the flip chip configuration as it is the emitting side. In some advanced chips, the sapphire is removed using a laser lift off technique (thin film flip technology: see Chapter 4).

Chips are then singulated from the wafer using the standard diamond sawing or laser dicing procedure. However, sapphire is very hard and requires specific dicing conditions: high speed rotating blades and low advance speed of the chuck, high power lasers. To maximize light extraction, it is better to have chamfered edges on the sides of the chip. We will discuss this point again in section 5.4.

Packaging followed the evolution from regular LEDs to high power or High-Brightness LED components and is influenced by size but also by heat dissipation and light extraction.

	Standard LED	**High-Brightness LED**	**Ultra-High-Brightness LED**
Size (μm)	350x350	~ 1,000x1,000	Up to 2,000x2,000
Input	30 mA, 3.5 V	350 mA, 3.5 V	1 to 1.5 A, 3.5 V
Electrical power (W)	< 0.2	Up to 1	Up to 5 or 10
Luminous flux (lumen)	1 to 3	5 to 50	100 to 500
Efficiency (lm/W)	20	> 50	> 100

Table 5.1. *Evolution of the main characteristics of LED chips influencing packaging*

The following section discusses the main types of LED packaging.

5.2.3. *Components with connection pins*

The majority of these traditional components with pins use classic lateral or vertical configuration chips (as in the following example). A figure showing such a component is presented below.

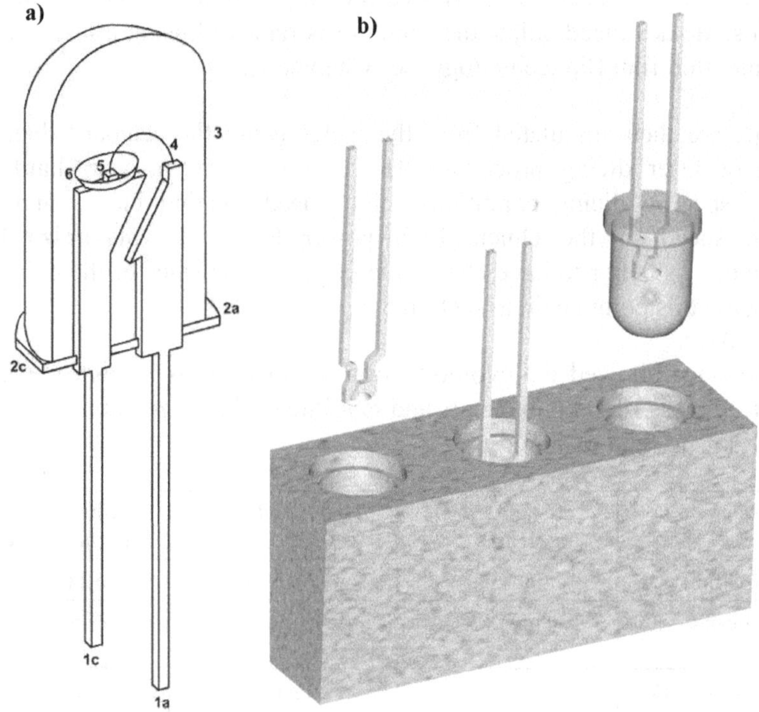

Figure 5.4. *(a) Schematic representation of a traditional LED with pins [MON 07] and (b) its fabrication process by injection and molding*

The flip side of the chip (5) is glued or soldered in a metallic cup (6) and electrically and thermally connected to one of the connection pins (1c). The upper side of the chip is connected by a conductive metallic wire, generally gold (4), to the other pin (1a). The system is encapsulated in a protective plastic epoxy resin using an injection molding technique shown in Figure 5.4b to ensure a mechanical protection against humidity and all other chemical products used for the mounting of the component (solder flux, cleaning agents, etc.). This encapsulant also works as a lens to spatially shape the emitted light beam. Finally, this lens can be transparent or colored depending on the application, but in most cases, these conventional components do not include a phosphor conversion material. Section 5.4 will discuss the different types of encapsulants. The metallic cup is most of the time covered with a reflective silver coating to enhance light extraction towards the top of the component. The pins are enlarged close to the cup to improve heat dissipation.

5.2.4. *SMT leadform components*

The wide variety of leadform components is often specific to each manufacturer. We will limit ourselves to discuss the main fabrication steps. Such a component roughly consists in six main elements: a mechanical support acting as an electrical contact and a heat sink [1] on which the chip is attached [2], an optional phosphor material for light conversion [3], an encapsulant [4] to protect the chip and fix a lens type primary optics [5] that shapes the light beam. Metallic wires (not shown) connect the chip's polarities to external interconnection pins [6]. These pins allow further connection and mounting of the component on the motherboard's surface.

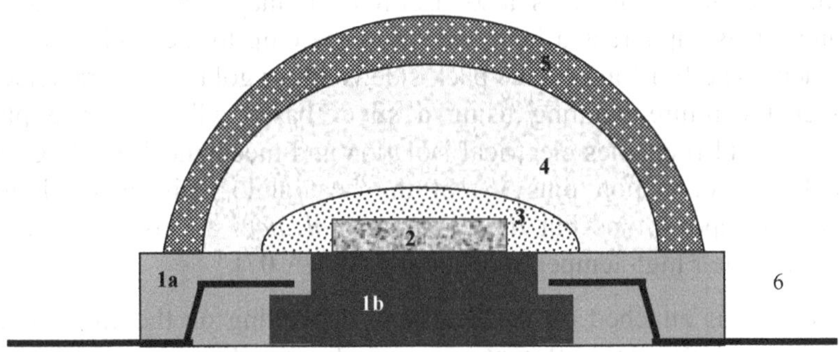

Figure 5.5. *Representation of a white LED made of an interconnection support and a heat sink [1], a LED chip [2], phosphor conversion material [3], an encapsulant [4] a lens [5] and interconnection pins for further mounting on the motherboard [6]*

For monochrome red, green or blue LEDs, there is no phosphor conversion material; the desired color is generally obtained directly at the chip level. For the so-called white LEDs, two approaches are possible: "RGB" (red, green, blue) and pc-LED (*Phosphor Converted Light-Emitting*

Diode). In the first case, the white light can be obtained by using the three chips (red, green and blue) in a single component. In the second case, a phosphor conversion material is deposited on the emitting face of one or more chips integrated into the component. This technique is used for instance in very high performance diodes where several chips are integrated in a single component to yield Ultra High-Brightness LEDs having a very high luminous output per package. The LEDs are first glued or soldered to an interposer including an interconnection network, then interconnected in series using gold wires. The interposer is then attached on the component's base and another wiring enables interconnection from the interposer to external interconnection pins:

– the base (1) has a mechanical function as well as a heat dissipation function. It is made of a heat sink (1b) molded in an insulated polymer (1a) with external interconnection pins (6). The heat sink is normally made of copper but other metals like aluminum can be used for cost reduction purposes. The upper surface of the heat sink where the chip is attached is generally coated with a thin layer of silver that reflects light to enhance light output towards the top of the component. This surface can be plane or concave to shape the light beam according to the desired spatial radiation pattern. A concave shape is also useful as it may serve as a phosphor container. This support is a heat sink from the chip to the back side of the component. The finishing on the back side is either gold for future soldering or silver for future bonding using a silver based adhesive. The plastic overmolding (1a) enables electrical isolation and mechanical bonding of the external interconnection pins (6) of the heat sink. This overmolding is generally an epoxy type but *liquid crystal polymers* are used increasingly often due to their high temperature stability [KRA 07];

– the chip is attached on the heat sink depending on the chip structure (vertical or lateral) using different processes discussed later on;

– gold wiring (not shown) enables interconnections between the chip and the interconnection pins;

– once the chip is electrically connected, the phosphor material made from an inorganic powder and a polymer binder (a silicone gel most of the time) is deposited on the chip. Depending on the method used, the product's viscosity and the metallic base's shape, the deposition layer's thickness varies from a few dozen to hundreds of micrometers;

– the lens shown (5) has a semi-spherical hollow shape, but many variations of the system exist (depending on the desired spatial radiation pattern); in the example presented, the polymer lens is simply deposited and an injection molding process allows the encapsulant (4) to be injected right above the chip as a protection and a refraction index matching. The polymer lens is generally made of epoxy, but silicone is being used more often for its better stability in high temperature and UV radiation environment (see section 5.4). The use of a unique material for the lens and encapsulant simplifies the overall manufacturing process since both two parts become one, obtained by a unique injection molding process.

5.2.4.1. *Vertical LED chip integration*

In this configuration, the whole back side of the chip is generally the anode and the front emitting side presents a grid or simple cathode in the case of more conventional or size reduced LED. The LED chip is soldered (Au-Sn alloy most of the time) or glued (using an epoxy silver based paste) on the heat sink.

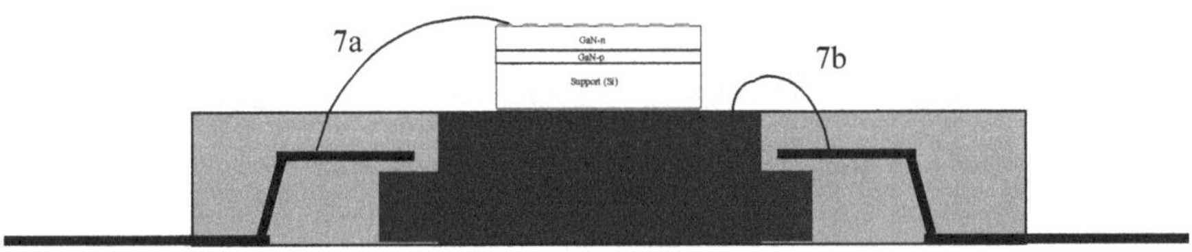

Figure 5.6. *Vertical LED chip integration*

The benefit of the soldering option is that the thermal conductivity of the interface is higher than that of paste interface. However void-free soldered joints are still challenges, especially without deoxidizing flux. In fact, the widely used flux could possibly damage the LED chip; therefore, other procedures are used like complex flux-less Au-Sn soldering processes, under a reducing atmosphere for instance. More details concerning conductive adhesive can be found in the following sources [LIC 05, LIU 99] as for brazing alloys in [JAC 93]. One or more Au wires (7b) connect the upper part of the heat sink to one of the interconnection pins forming the anode.

Other Au wires (7a) connect the front face of the LED chip to the other interconnection pin to form the cathode. In this type of packaging, the heat sink is electrically connected to the anode. Even though polarizing the heat sink can be a limiting factor for some applications, heat dissipation may be enhanced with this configuration. To avoid polarizing the slug, an electrical insulator but thermally conductive layer/interposer on which wiring can be done (7b), is placed between the heat sink and the chip.

A protective die against ElectroStatic Discharge (ESD) made of Zener diodes is sometimes placed on the heat sink and connected in parallel with the LED chip. Alternatively, Zener diodes can also be directly formed onto the insulator/interposer layer discussed previously.

5.2.4.2. *Standard lateral chip LED integration*

For the standard lateral chip LED, electrical contacts are on the same side. An integration similar to the previous one is possible for the standard lateral chip LED with an emission through the 2 "+" and "-" contacts.

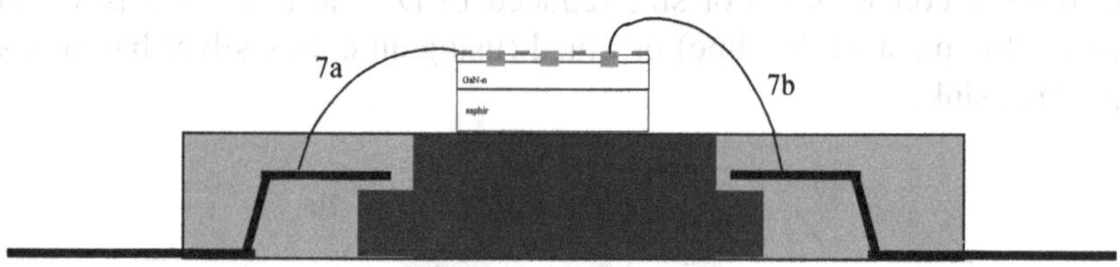

Figure 5.7. *Vertical LED chip integration*

In this configuration, the contacts create large shadowed zones very unfavorable to light emission. In this widely used implementation, the chip is glued or soldered on the heat sink but the sapphire from the bottom face of the chip allows the electrical insulation from the former. Au wiring is used to connect the top face contacts to external interconnections pins.

5.2.4.3. *Lateral flip chip LED integration*

In a flip chip configuration, the chip's "+" and "-" contacts zones are transferred on an interposer (8) using a flip chip technique with conductive glue or Au stud bumps. A polymer encapsulation material (underfill) fills the empty spaces between the stud bumps to enhance heat dissipation. This polymer is generally an epoxy adhesive filled with submicron silica particles to reduce the Coefficient of Thermal Expansion (CTE) and other ceramic submicron particles to improve thermal conductivity. As previously, the interposer includes Zener diodes for ESD protection.

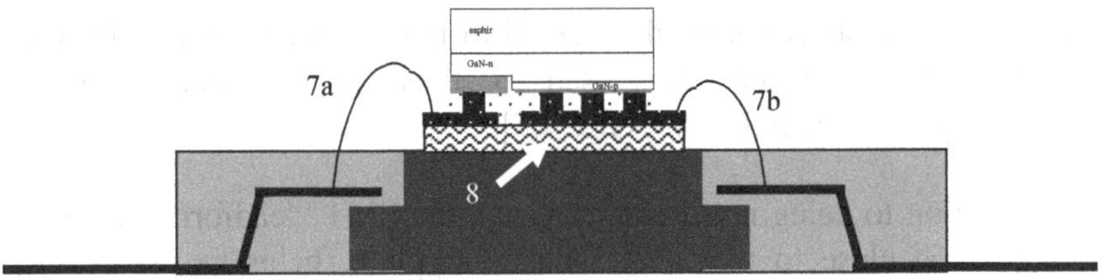

Figure 5.8. *Lateral flip chip LED integration [KRA 07]*

5.2.5. *SMT "leadless" components*

In this configuration, the chip's support does not integrate external connection pins and these interconnections are made through connecting vias. The system is detailed in the following figure.

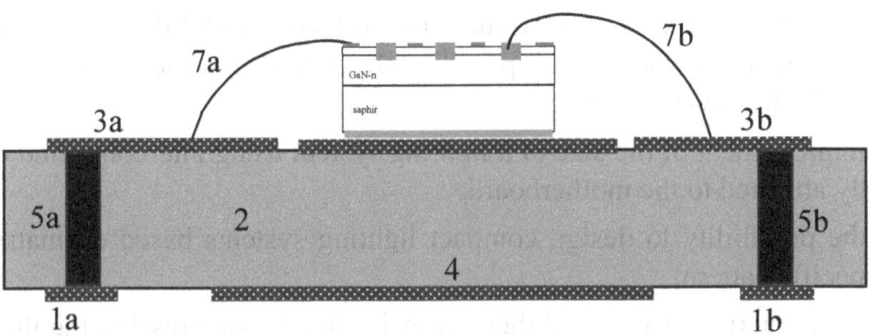

Figure 5.9. *SMT "leadless" component with ceramic base*

In this figure, the LED configuration is standard lateral but other types of LED can be packaged in a similar fashion to SMT "leadform" components. The support (2) is generally a ceramic allowing electrical insulation of polarities 1a) and b) and simultaneously ensuring a heat transfer from the chip to the back side of the component. The ceramic is in general alumina but other types of ceramics can be used like aluminum nitride (AlN), which has a better thermal conductivity. Surface interconnections (3) and through vias (5) are thick Cu (50 to 100 µm), Cu being a good electrical and thermal conductor. These interconnections can be coated with a thin layer of reflecting Ag around the chip to optimize light extraction on the top face of the component. On the component's back side, in addition to the polarities 1a) and b), a large Cu metallized zone (4) is soldered or glued to the motherboard to enhance heat dissipation. Ceramic substrates used for these components are of "direct bonding copper" type [CUR 07, GLO 07].

In addition to being more compact than the SMT "leadform", using a low CTE ceramic close to that of GaN and sapphire, helps to reduce thermo-mechanical constraints created by different thermal expansion mismatch stresses (see section 5.7.1). Nevertheless, using ceramics in such a component seems more expensive. Finally, we insist on the fact that the thermal conductivity of ceramics is lower than that of copper and therefore surface spreading is required (see Figure 5.12).

5.2.6. *Other technologies*

5.2.6.1. *Chip-on-board technologies*

This packaging process fits right between level 1 and 2 defined in section 5.1. As a matter of fact, the chip is directly attached on the motherboard to address the following purposes:

– improvement of the size of a lighting system using interconnected chips directly attached to the motherboard;

– the possibility to design compact lighting systems based on matrix or any specific pattern;

– the reduction of the total thermal resistance by suppressing the thermal resistance of the components.

The motherboard is generally made of various kind of substrates:

– polymer: the most famous one is glass fiber reinforced epoxy (or FR4);

– metal: these substrates are called IMSs (*insulated metal substrates*) or MCPCB (*metal core printed circuit board*) and are made from a metal core (which is generally a low CTE material made of Kovar or Invar type, aluminum alloy or even copper (depending on the power to dissipate) coated with a thin polymer electrical insulation layer (polyimide, benzocyclobutene, etc.); This layer is also generally filled with inorganic particles to increase thermal conductivity and reduce the CTE; finally, a copper wiring network enables interconnection with the chip;

– ceramics: these are DBC substrates, as described in the previous section.

Polymer substrates are generally used for common low cost, mid-luminous power applications while metallic and ceramic substrates are used in high luminous power applications with high power LEDs like in the automotive or general lighting industry for instance. Indeed, in these last applications, heat dissipation needs and CTE mismatch between chip and substrate may become critical.

5.2.6.2. *Silicon packaging*

Various other LED packaging method exist. Let us mention an example of a silicon carrier developed by Hymite [HYM 07] using traditional silicon processing methods used in microelectromechanical systems (MEMS) fabrication.

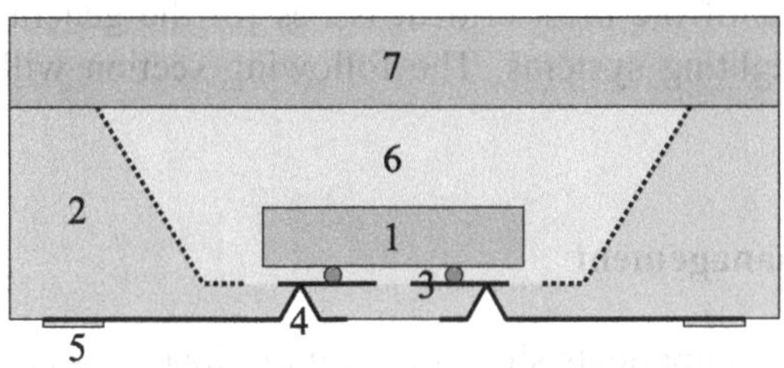

Figure 5.10. *LED silicon carrier [HYM 07, MUR 07]*

Many carriers are manufactured collectively on a common silicon wafer (2) using anisotropic etching of cavities allowing the chip's placement (1) and serve as a receptacle for phosphor and encapsulant materials (6). In the previous figure, the chip is interconnected using the flip chip method (3) in the Si cavity, but a gold wiring bonding can be used for lateral and vertical LED configurations. The interconnections between the chip and the back side of the SMT component are obtained using metallic vias (4). Bumps (5) are present for soldering on the motherboard. The component is protected thanks to a simple transparent protective lid/cap or lens (7) placed on top of the carrier. This type of packaging gains from the correct thermal conductivity of silicon and a collective fabrication process. The edges of cavities can be metallized to improve light extraction. Primary optic systems can be realized collectively at the wafer scale by overmolding polymer materials.

5.2.7. *Conclusion*

The technologies listed below are not exhaustive. However, the main criteria determining the choice of a LED packaging solution should be:

– the lateral or vertical structure of the LED chip;

– the electrical, thermal and luminous output power of the LED chip;

– size and environment constraints which can be very different depending on the targeted application, for instance, a mobile phone versus a car's headlight.

Industrialization constraints, depending on the application, often lead to a tradeoff between performance and cost of the component. Packaging evolution must deal with a rapid increase in brightness and heat dissipation now becomes one of the most critical issues for the generalization of LED components in lighting systems. The following section will discuss this in more detail.

5.3. Thermal management

All electronic components share the heat dissipation issue. [HAR 04, PET 01] cover this problem in more detail. We will limit ourselves to recalling some basics: application specificities of LEDs compared to other electronic components and a thermal comparison between the several packaging processes described in the previous section.

5.3.1. *Motivations*

Nowadays, the highest performance LED components still have a limited total efficiency since about 20 to 25% of the injected electrical power is converted in visible light, the rest being lost as heat dissipation [KRA 07, ZEH 07]. Typically, thermal power densities reach at the chip level 100 to 200 W/cm^2, which classifies LEDs as true power electronics components. Without proper thermal management, the component's local heating has several harmful effects that affect the LED's integrity and performance:

– the heating of the component induces internal and external thermo-mechanical stresses due to the CTE mismatch and can create cracks, delaminations and other physical damage. Indeed, as for all electronic components, the lifetime is directly impacted by repeated thermal cycling [HAR 04];

– the heating of the component also leads to an accelerated aging of the LED materials, especially polymer materials. Optical properties of the encapsulants and lenses are also affected and become progressively opaque;

– a rise in the LED's junction temperature leads to a decrease in the LED's internal quantum efficiency and thus in the luminous output. Moreover, for white pc-LEDs, the phosphor's conversion yield is also reduced with an increase in temperature;

– finally, a temperature rise in the junction active area creates inter-diffusion and solid state defects that, at a constant power, decrease the lifetime of the LED depending on the semiconductor used in the considered diode. Typically, a reduction by a factor of 2 in the lifetime of the component is observed for a junction temperature rise of about 15°C [LUM 07].

5.3.2. *Heat dissipation modes*

In a LED lighting system, the induced thermal power can be dissipated traditionally through the three conduction convection and radiation modes. The thermal flux can be expressed as follows:

$$P_d = \frac{T_j - T_a}{R_{th}} \qquad [5.1]$$

where P_d is the generated thermal power in Watts, T_j is the LED's junction temperature in the active area (see also Chapter 6), T_a is the ambient temperature and R_{th} the total thermal resistance of the system (expressed in Kelvin/watt) including both the component and the motherboard.

The conduction mode's contribution to the thermal resistance in a simple case for a sheet of thickness L and surface S is as follows:

$$R_{th\,(conductive)} = \frac{L}{K_T \cdot S} \qquad [5.2]$$

where K_T is the thermal conductivity of the sheet's material in W/m.K.

The convection mode's contribution to the thermal resistance is the following:

$$R_{th(convective)} = \frac{1}{h_c \cdot S} \qquad [5.3]$$

where h_c is the thermal convection transfer coefficient; h_c depends on air flow conditions at the vicinity of the thermal exchange surface S (free flow convection in laminar regime, forced flow convection, etc.); in free flow convection, h_c depends on the vertical or horizontal position of the considered surface; it is generally determined using an empirical method.

Finally the radiation's contribution to the thermal resistance is "simply" expressed as:

$$R_{th(radiative)} = \frac{1}{h_r} \qquad [5.4]$$

where h_r is the radiative exchange coefficient; h_r is proportional to the exchange surface area and the material's emissivity and depends mainly on the emissive body's temperature, since it is a function to the 4th power of the temperature of the radiant black body.

As a first approach, we can define and thermally design a system using the following equations and mathematical equations from thermal resistance calculations found in [HAR 04, PET 01] applying the Ohm's law equivalence as follows:

$$\Delta T = R_{th} . P_d \Leftrightarrow U = R . I \qquad [5.5]$$

Mainly, we can observe that thermal resistances in series add and the inverse of the thermal resistances add for resistances in parallel, just like electrical resistances. From the formulae cited above, we will focus on the following principles:

– thermal dissipation through conduction is of course a function of the internal thermal conductivity of used materials; a non-exhaustive list of physical properties of materials used in packaging can be found in section 5.7.1;

– thermal dissipation is proportional to the exchange surface, whatever the thermal transfer mode considered. It is thus clear that for a packaged LED component where the size is fairly small, only conduction is considered, while convection and radiation mainly occur at the board and system level.

5.3.3. *Thermal dissipation in LEDs*

To achieve the thermal design of a lighting system, several simulation tools are available: ANSYS, FLOWTERM, COMSOL, etc. We will not provide more details on them since they are not specific to thermal dissipation in LEDs and would constitute a chapter on their own.

It is also possible to qualitatively and macroscopically characterize heat dissipation in lighting systems using an infrared camera. This method has the advantage of qualitatively or semi-qualitatively comparing different packaging modes by localizing the main heating zones.

In this section, we will give a few simple thermal characteristics of power components applicable to LEDs. These characteristics can qualitatively explain heat dissipation phenomena and assist in a first simple design dimensioning. As specified in the previous section, the thermal resistance $R_{j\text{-}a}$ (in K/W) of a lighting system is defined by the following equation:

$$R_{j-a} = \frac{T_j - T_a}{P_d} \qquad [5.6]$$

where T_j is the p-n junction's temperature, T_a is the ambient temperature and P_d is the dissipated thermal power. For a simple modeling of an LED's thermal dissipation, we consider that all the heat produced at the junction is dissipated through the bottom side of the chip.

In fact, it is the desired goal, since an important increase in temperature would lead to a major decrease of the phosphor conversion material efficiency and a premature aging of the polymer and encapsulant of the lens.

The thermal power generated at the junction (T_j) is first dissipated through conduction in the chip's substrate, then through the joint on the support and then through the slug (T_s). The heat is then dissipated through the joint on the motherboard (T_b) still through conduction, and finally through convection and radiation from the motherboard and system to the ambient air (T_a), as illustrated in the figure below.

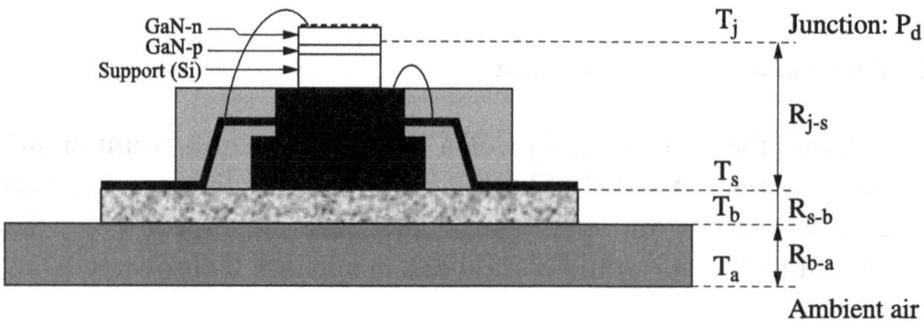

Figure 5.11. *Schematic modeling of the thermal dissipation of a LED attached on a motherboard*

The thermal resistance $R_{j\text{-}a}$ is the sum of the resistances $R_{j\text{-}s}$, $R_{s\text{-}b}$, $R_{b\text{-}a}$. The LED component (level 1 packaging) is thus often defined by $R_{j\text{-}s}$. The contribution of the conduction $R_{j\text{-}s}$ mode can be approximately determined analytically using formula [5.2] and knowing the packaging materials, their dimensions and thermal properties. It can also be determined experimentally by maintaining the ambient temperature at the bottom of the slug while measuring simultaneously the junction's temperature and the dissipated power. The junction's temperature can be found either by measuring the primary emission peak of the diode or by measuring the direct voltage and referring it to a previously performed temperature calibration [SCH 04]. Through approximation and knowing that about 80% of the injected electrical power is dissipated as heat, we will assume for calculations that the thermal dissipated power (P_d) is equal to the electrical power injected, the electrical power being the product of the current and the voltage applied to the LED. Conversely, equation [5.1] can be used to find the junction's temperature in another way, by knowing the thermal resistance of the slug, the value of $R_{j\text{-}s}$ being given in the technical data sheet of the LED's manufacturer.

For a lighting system with a given efficiency, the maximum luminous output is limited by the maximum electrical power that can be injected in the component. For reliability and performance reasons, the maximum junction's temperature in today's LED components is typically around 150°C to 180°C. Using equation [5.1], we clearly see that the maximum electrical power, and therefore the maximum luminous or luminous output power, is directly dependent on the total thermal resistance $R_{j\text{-}a}$ and thus for the component only on the component thermal resistance $R_{j\text{-}s}$. As a result, we see that it is possible to maximize the optical power emitted by enhancing the thermal dissipation and reducing $R_{j\text{-}s}$. However, gains made on the thermal resistance will always be compromises with related costs associated with the optimization of the thermal dissipation. Furthermore, apart from a few specific high power lighting system applications, it seems unrealistic to use active dissipation systems or very large dissipation systems for cost and size issues at the component level.

5.3.4. *Comparison of different packaging processes*

The knowledge of thermal dissipation mechanisms explained in the previous sections shows that the evolution of LED packaging has addressed a major reduction of the components thermal resistances from ~ 250 K/W for pin components to ~ 15 K/W for "leadform" components, to even less than 10 or 5 K/W for "leadless" components [ARI 02]. To obtain such a small thermal resistance, LED manufacturers worked on several points.

Interconnections

First of all, internal interconnections within the chips and components are optimized to reduce electrical resistance and thus limit heating by acting on several points: the chip's internal electrical resistance interconnection reduction, the increase of the number of Au wires and their cross-sections, as well as the number of external interconnection pins and use of low resistivity and high thermal conductivity materials like Au and Cu for these interconnections. A non-exhaustive list of physical properties of materials used in packaging can be found in section 5.7.1.

Bonding materials used on the support

In a few rare cases, components can be mechanically maintained on the support with a pure metallic thermal interface material (like In for instance) or polymeric (for instance, thermal grease filled with inorganic or metallic particles to enhance thermal conductivity). In most cases, the bonding material is a conducting glue or a metallic solder alloy. There are a wide variety of thermally conductive glues with a polymeric base, generally epoxy, but it can also be a silicone or an acrylic, etc. In addition, an inorganic (AlN, SiC, BN) or a metallic filler (Ag, Ni, etc.) is added. In the inorganic filler case, the adhesive is a strictly thermally conductive, whereas in the case of a metallic charge, it is an electrical and thermal charge. Today, in the state of the art, common inorganic filled adhesives have a maximum conductivity of a few W/m.K up to a few dozen W/m.K for those with a metallic filler. These last epoxies filled with micrometric and nanometric Ag particles may have a thermal conductivity up to 30 W/m.K which is outstanding for adhesives [EPO 07]. However, in the future, filled adhesives with carbon nanotubes or diamond with a better conductivity will be available [ZWE 06]. More details about different adhesives can be found in [LIC 05, LIU 93]. We can also observe that Ag-based adhesive have a very high reflective power, which can be a great advantage for light output. Concerning solder alloys, their thermal conductivity is generally much higher, typically 30 to 60 W/m.K (see section 5.7.1). The most common solder alloy for support attachment is the Au-Sn alloy because it has a good thermal conductivity and a higher melting point than other standard lead-free alloys (Sn-Ag, Sn-Ag-Cu, etc.) traditionally used to mount components on the motherboard. Besides, and in some conditions, the Au-Sn alloy can be implemented without flux, which limits contamination risks for the chip during mounting operations. In the case of soldering, the alloy is in general deposited on the back side of the chips at the wafer scale using thin layer deposition techniques (see section 5.2.2).

Substrate materials

For "leadform" components with a metallic base, copper is currently the most frequently used because it is the metal that has by far the highest thermal conductivity (~ 400 W/m.K). Au and Ag also have a very good thermal conductivity but are much more expensive and do not have a mechanical strength as good as Cu. Al is also widely used for cost reasons, even though its thermal conductivity is lower.

For "leadless" components, the most frequently used ceramic is alumina, even though its thermal conductivity is rather poor, because it is widely used in electronics (especially power electronics) and its cost is moderate. AlN has a thermal conductivity typically five to ten times higher than alumina, but it is still infrequently used for cost reasons. It is highly probable that if the LED market develops, AlN, which is widely used in DBC substrates, will replace alumina for power electronics components. Many ceramics like beryllium oxide (BeO), cubic bore nitride (CBN) or diamond have an excellent thermal conductivity, but their use is again limited for cost reasons.

Ceramics often have a low CTE close to the semiconductor constituting the diode (GaN for example) and the growth substrate (sapphire) which, unlike metals, can be critical for very high brightness LED packaging with sizes goes from 350x350 μm^2 to 2x2 mm^2. Finally, composite materials like Cu-Mo and Al-SiC are a good compromise in terms of CTE and thermal conductivity.

Concerning *chip-on-board* systems, as described in section 5.2.6.1, with the increase in power of LEDs, substrates evolved from FR4 polymers to insulated metal substrates (IMS or MCPCB) and DBC too. IMS's advantage, compared to DBC substrates, is mainly cost when compared to ceramics. The drawback of IMS is its higher CTE than most ceramics, but on the other hand, the insulation layer of the metal core remains a thermally resistive layer even if the thermal conductivity is enhanced by the addition of highly conductive inorganic particles, AlN, BN, SiC, etc.

Design

It seems clear, when looking at the expectations in terms of LED compactness, that any surface extension is limited. In order to reduce thermal resistance, chips are thinned on the back side and on the emissive side, whether they are Si or Ge for vertical or sapphire for lateral LEDs, to a typical thickness of 100 µm. It is highly likely that, in the near future, the base substrate of the chip will be thinned down to less than 50 or even 25 µm. We may also add that Si and Ge substrate vertical LED chips constitute a very good choice for heat dissipation optimization compared to standard lateral LED chips.

Concerning lateral flip chip LED configuration, the chip's substrate is on the opposite side of the chip to the heat sink, which sounds attractive in terms of heat dissipation. However, the current method used to attach chips, using Au stud bumps, limits the thermal conduction exchange surface, even though Au has a great thermal conductivity and the empty space between the stud bumps is filled by a thermally conductive adhesive. Since the "+" and "-" contacts must be electrically insulated, the use of a thermally conductive but electrically insulating adhesive is necessary despite the thermal conductivity being lower than a thermally and electrically conductive adhesive. Another drawback for heat dissipation is the use of an interposer to ensure the interconnection and the electrical insulation of the metallic base. A "leadless" carrier configuration seems more adapted for flip chip lateral LED systems. Theoretically, it is possible to enhance heat dissipation by replacing the Au stud bumps used currently with a full contact surface solder. This solution still raises a thermo-mechanical problem due to the proximity of the active zone with the solder joint on the base. As for LED chips, the thickness of DBC SMT components tends to decrease from 1 mm in current standards to less than 500 μm.

The enhancement of heat transfer (i.e. conduction, radiation and convection) in a lighting system can be illustrated in the following figure through the concept of thermal spreading, whether it is the component's integration at the motherboard level or the chip at the component level or even in chip-on-board modules. A first thermal path enables the dissipation of heat through conduction from a chip or LED component (1) linked through soldering or gluing on a substrate (2) made of an electrically insulating material on which a thermally conductive layer (3a) with a large exchange surface area has been deposited. The thermal spreading occurs on the front side where the LED is placed; if the electrically insulating material (2) is a thermally conductive, the thermal spreading also occurs on the back side (3b). This technique can be used for an alumina/Cu DBC substrate.

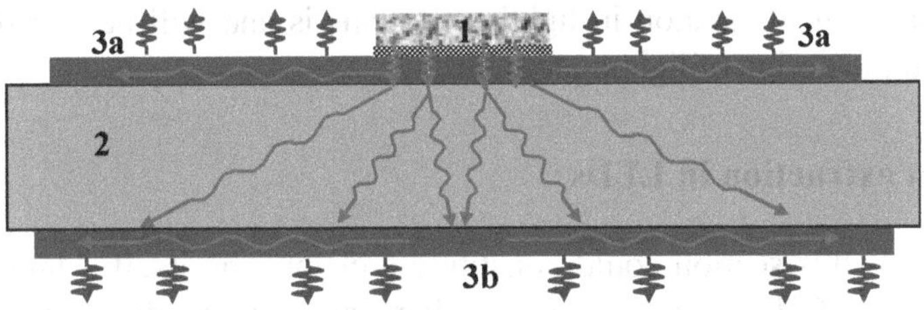

Figure 5.12. *Schematic representation of thermal spreading through conduction and convection/radiation (see color plate section)*

Finally, it is possible to optimize heat dissipation by using injection molding of polymers with a better thermal conductivity. An inorganic filler with AlN, SiC, BN, etc., is thus used, as for thermally conductive adhesives.

5.3.5. *Conclusion*

In this section, after summarizing the motivations for thermal management study in LED components, we recalled different heat transfer modes and their specificities in LEDs with basic information on the junction's temperature and thermal resistance. We performed a confrontation analysis of several existing carriers by showing all the possible evolutions and optimizations and their packaging processes. Most of the time, it is clear that using the existing high performance heat dissipation method is often very expensive. It is probable that the use of materials with high thermal conductivity will develop with increasing production volumes.

In the analysis, we considered that heat was only generated at the junction. As a matter of fact, a deeper analysis of the total luminous output of the most performing High-Brightness LEDs shows that the generated heat is also located at: a) the chip's p and n internal interconnections (20%), b) the chip's primary radiation absorption (5%), c) non-radiative emission (40%). Only 35% of the injected power is converted into light radiation. Of this 35%, only about 20% is converted into visible light due to losses in primary light phosphor conversion into visible radiations [KRA 07].

In the near future, thanks to several improvements in the LED's structure, the phosphor conversion, the packaging, etc., about 50% of the injected electrical power could be converted into visible light [KRA 07, ZEH 07]. It is obvious that heat dissipation in lighting systems is and will be a critical and key issue in the design of lighting systems.

5.4. Light extraction in LEDs

As before, this section could constitute a chapter by itself. Many details are given in [SCH 06] for instance. There are several different ways to maximize light extraction like structuring the LED chip's surface, as discussed in Chapter 4. Light extraction improvements can also be achieved through an optimization of phosphor conversion material, as shown in Chapter 4. In this chapter, we will limit ourselves to recalling the main important optical aspects and light extraction concepts related to packaging.

5.4.1. *Lateral light extraction in LEDs*

Light radiation created at the junction level must be extracted from the semiconductor or the sapphire depending on the vertical or lateral structure of the diode.

Light extraction from a source (S) to an outside connected environment (A) is ruled by a first phenomenon called critical angle and following the equation:

$$\frac{n_A}{n_S} = \frac{\sin \varphi_S}{\sin \varphi_A} \qquad [5.7]$$

where n_A and n_S are respectively the refractive index of environment A and of the semiconductor S, Φ_S is the incident angle at the interface between the two media and Φ_A is the refracted angle in the surrounding environment A. The critical angle of the total reflection Φ_C is obtained when the refracted angle in environment A is equal to 90°.

In the case where the extraction environment is air, knowing that the LED's semiconductors are high index materials, the internal refractive critical angle leads to multiple total reflections inside the constitutive material of the LED. This results in poor light extraction efficiency. For example, GaN's refractive index is 2.5 with a critical angle of 24°; as a consequence, the light extraction cone is significantly reduced. To overcome this effect, LEDs are generally shaped on their edges into shaped cut pyramids [KRA 99] to reduce the mean free path into the diode and to favor lateral light extraction, as shown on the following figure.

An increase in the extraction by a factor of 2 is possible by using this method. The laterally extracted light is then reflected outside the chip on a metallic surface whose refraction index is the lowest possible. Most efficient materials are respectively Ag, Au and Al.

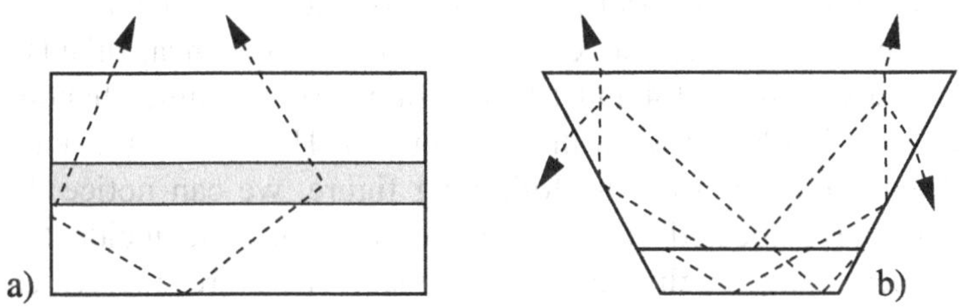

Figure 5.13. *LED's light extraction principle with (a) vertical edges, (b) pyramidal truncated shaped edges enhancing lateral light extraction [KRA 99]*

5.4.2. *Vertical light extraction through a lens*

In order to optimize light extraction on the upper side of the chip, the surface is structured, metallic reflectors are inserted or even more complex processes in the diode itself are implemented efficiently (see Chapter 4). These improvements and the increase in the surface area of LEDs have rendered the need for truncating processes shown in the previous section useless. In this section, direct ways to extract the light vertical through the emissive side are discussed. Using equation [5.7] and assuming the light source to be a point, it is possible to calculate the ratio between the extracted power (P_A) in an environment A, and the source power generated in the diode (Ps) at the interface between the two media:

$$\frac{P_A}{P_S} \sim \frac{1}{4} \cdot \frac{n_A^2}{n_s^2} \quad [5.8]$$

To simplify the following analysis taken from [NUE 69, SCH 06], we will consider the simple case where the encapsulant and the lens are unique materials. Equation [5.8] shows that the light extraction yield is directly governed by the refractive index of the environment where the light is extracted. Most of the semiconductors of the LEDs have a high refractive index. For GaN LEDs, for instance, the refractive index is 2.5, which leads to a critical angle of about 24° and a loss in extracted light power of a factor of 6.25 for an extraction into air. By using an encapsulant with an refractive index of 1.5, the critical angle becomes 37°, which dramatically increases the emission cone, and the loss in light power is only of a factor of 2.8, compared to an extraction performed directly into air. Most of the time, for encapsulant and lens materials, we are looking for optically transparent materials with a high refractive index. Most of the encapsulants have a refractive index between 1.4 and 1.8. The same critical angle phenomenon is observed at the border between the lens and air. However, by using a semi-spherical lens, as shown on the following figure, we can notice that for a point light source located at the center of the lens, the incident angle is always perpendicular to the air/lens interface. Therefore, the critical angle Φ_C leading to a total reflection is never reached. In the case of a physical extended light source, there are a few angular configurations for which there is a total internal reflection in the lens, but the overall extraction remains much better than for a planar lens.

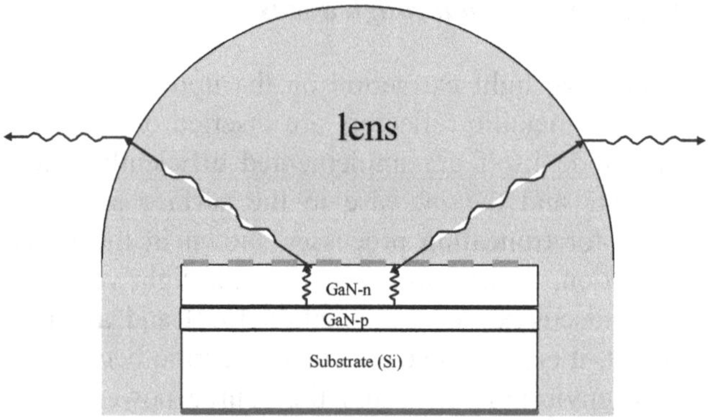

Figure 5.14. *Emission configuration through a semi-spherical lens having an intermediate refractive index between those of the semiconductor and air*

As for the chip/lens interface, the light extraction efficiency at the lens/air interface is directly connected to the difference in refractive index as in formula [5.8]. The use of a semi-spherical lens with a high refractive index has two advantages: 1) it maximizes light extraction at the chip/lens interface and 2) it makes it possible to spatially spread the emitted beam, and avoids or limits total reflections at the lens/air interface. It is possible to modify the spatial radiation pattern by modifying the shape of the light source from a planar to a curved mode. The emission beam is called "Lambertian" in the case of a planar source, whereas a very curved LED surface leads to an increasingly focused light beam. In general, the LED surface is always planar (and/or structured) and the shape of the light beam is created by modifying the shape of the lens instead of the chip's surface.

For lateral configuration flip chip LEDs, the light radiations are first extracted from the GaN into the sapphire and then in the encapsulant and/or the lens. The refractive index of sapphire being 1.77, the same critical angle phenomenon is observed. Moreover, sapphire is already an intermediate refractive index material between the GaN and the encapsulant and thus helps in extracting the light. However, since sapphire substrates are always planar, the incident internal critical angle of total reflection is often reached, which limits the efficiency of the extraction in addition to internal absorption by sapphire.

Finally, in addition to surface structuring, we can notice that the use of multi-layers with refractive index gradients can optimize light extraction at the surface of the chip. These layers, also called "anti-reflective coatings", are made of variable refractive index materials. We can name, for instance, the most well known: ZnS (2.3), silicon nitride (index = 2), alumina (1.76) and silica (1.45).

5.4.3. *Lens/encapsulant materials*

Historically, a LED chip was encapsulated and thus protected mechanically and chemically (against humidity especially) and the encapsulant would also play an optical role as shown above; the lens was sometimes away from the encapsulant without optical coupling and played only as a light beam shaping. The two terms "lens" or "encapsulant" refer in principle to two different characteristics and functions. The "lens" is used to modify the spatial pattern beam and the "encapsulant" to protect the LED chip and enhance light extraction. In the following text, we may use either term interchangeably since, as mentioned previously, they may be a single part of unique material. As we discussed, from an optical point of view, the two main characteristics that are important for the choice of lens material are a high optical transparency and a high refractive index. Moreover, since the encapsulant is right above the chip and sometimes incorporates Au connection wires, it must be soft to balance and moderate CTE mismatch stresses. The fully packaged component will then be soldered on a motherboard at temperatures of about 200 to 250°C. It is very important that optical properties of the encapsulant are not affected during this mounting step. Finally, the encapsulant must withstand thermal aging (during the 100 to 150°C working conditions) and possible optical aging (continuous UV radiations from the environment). This thermal and optical aging leads to a progressive opacity of the lens. Detailed information about formulae and general characteristics can be found in [LIC 05]. To fulfill these criteria, the most well known polymers are epoxy, silicones and acrylics.

A cured epoxy resin is obtained using a reticulation polymerization process of a monomer base polymer with a generally aminic or acidic hardener (a network is created using polymerization) with at least two epoxy functions (see Figure 5.15), and whose chain length and number of functions can influence the final mechanical properties. This reaction is achieved with a reticulation hardener and optionally a catalyst accelerator:

$$\text{monomer} + \text{hardener} \rightarrow \text{cured epoxy resin (three-dimensional network)}$$

There are numerous hardeners that do not directly determine the final properties of the physical material but influence the curing conditions (time, temperature, etc.). The most frequently used hardeners are amine-based. Products are called bi-component when the epoxy monomer and the hardener are separated and polymerization occurs when they are mixed together. For single-component products, the mixing is done at the production step and is preferably stored at a low temperature (-20 to -40°C) to inhibit reaction before use. In both cases, the reaction is thermally activated.

Name	Formula
Epoxy function	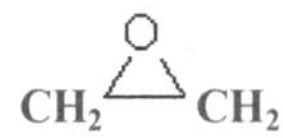
DGEBA = Poly(Bisphenol A-co-epichlorohydrin), glycidyl end-capped	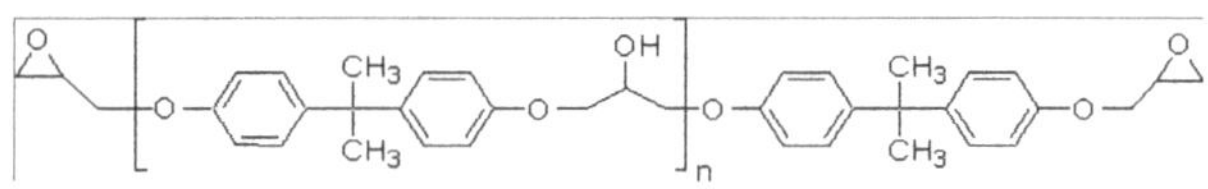
Epoxy Cresol Novolac = Poly[(o-cresyl glycidyl ether)-co-formaldehyde]	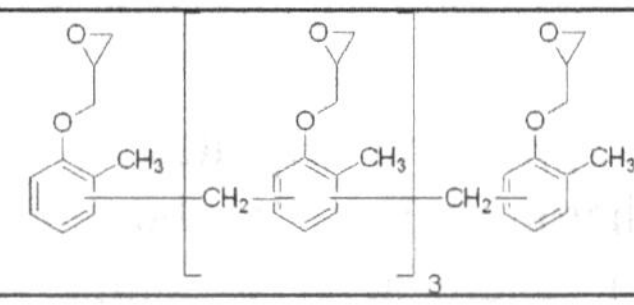
1,4 Butanediol diglycidyl ether	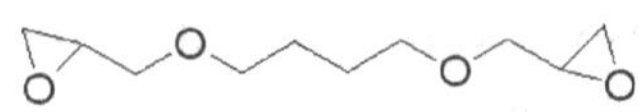
Resorcinol diglycidyl ether (RDGE)	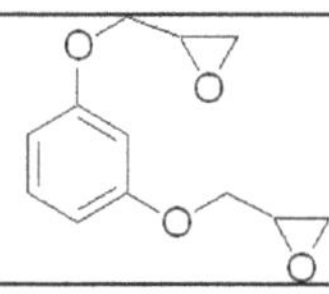

Figure 5.15. *Standard epoxy functions of a few components*

The mechanical properties of silicones are different from epoxies. These polymers are, unlike epoxies, composed of a network of Si-O bonds. Oddly, silicones are often very soft polymers in a wide range of temperatures, from -40 to 260°C typically, and have a very good stability in temperature. Finally, like epoxies, they have a good resistance against humidity. Silicone adhesives are generally divided into two categories: a) condensation single-component systems (called RTV (*room temperature vulcanizing*)); these systems cure in atmospheric humidity and release acetic acid during the reaction, b) bi-component systems catalyzed by platinum (Pt). Polymerization of these two categories of polymers is also thermally activated.

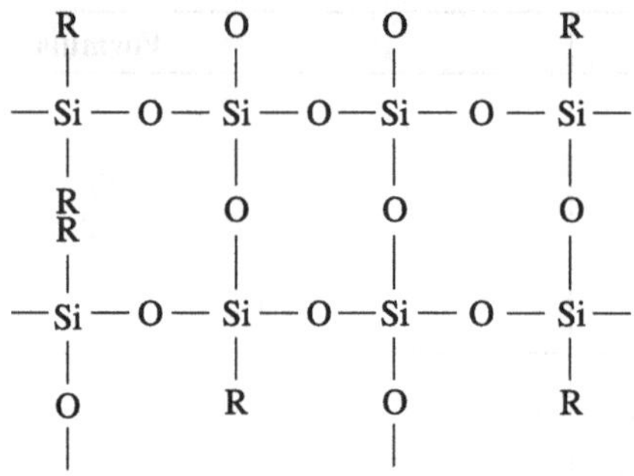

Figure 5.16. *Generic chemical formula of a silicone polymer*

Finally, acrylics and acrylates have mechanical properties close to those of epoxies (hardness D ~ 75), but curing times are generally very short (a few minutes). There are also single or bi-component versions and many acrylic adhesives can be cured at room temperature in a very short time using UV exposure.

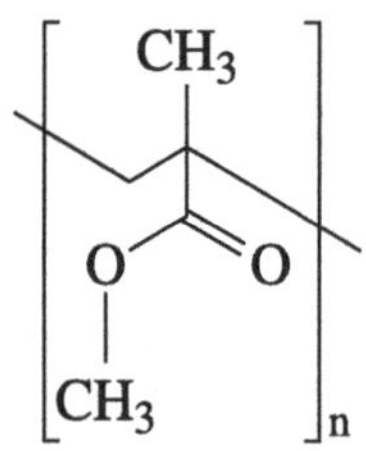

Figure 5.17. *Generic chemical formula of an acrylic polymer – polymethylmetacrylate*

Historically, the polymers most frequently used as encapsulants for LEDs have been epoxy materials because they possess all the previously cited criteria and were widely used in packaging and other applications, and were therefore relatively low cost polymers. Acrylics and acrylates are low cost polymers that can be cured at room temperature and/or under UV radiation, but like epoxies, their stability in temperature and UV exposure remains limited. With the generalization of High-Brightness LEDs, specifications like temperature and UV radiation stability are increasingly constraining and the use of silicones is now generalized in spite of its high cost and a moderate refractive index of about 1.6. For LED applications, silicones are often classified by their Shore hardness and are called gels or resins. Gels are extremely soft materials (Shore hardness A from 20 to 70) and are used as a matrix for phosphors and as encapsulants. Resins are harder materials (Shore hardness D ~ 50) and are generally used for lens fabrication; materials with intermediate properties also exist.

5.4.4. *Lenses and encapsulant implementation*

Materials used to encapsulate are available in gels, pastes or liquids of various viscosities. Lenses are generally created using specific injection-molding processes and can be shaped into any specific profile depending on the desired spatial radiation pattern. To facilitate lens manipulations, they can be mounted on metallic rings. Encapsulants are often implemented using direct deposition on the LED chips (or on the phosphor) previously placed on the carrier. Depending on the surface of the carrier (hollowed, planar, etc.), a viscous or even liquid product is used. There are a large number of implementation processes and their use depends on the viscosity of the product used. It is also possible in some cases to deposit using serigraphy or another thick layer deposition technique to obtain conformal encapsulants on the chip with an optimized thickness. Traditionally, the lens is implemented on the carrier; the encapsulant is then injected and molded to fill the cavity between the carrier and the lens. In the case where the lens and the encapsulant are one unique material, the polymer material is directly molded on the carrier.

5.5. LED component characteristics

In the previous sections, in addition to packaging processes, we discussed two critical points, heat dissipation and light extraction at the LED component's level. In this section, we will summarize thermal, optical and electrical characteristics of LED components that are important in the choice of a component and are specified in the component's technical data sheet.

5.5.1. *Thermal and electrical characteristics*

First of all, a LED is a diode and is described by an operating curve of current versus an applied voltage, I(V). There is a wide discrepancy of I(V) curves after manufacturing due to intrinsic properties of the junction, the packaging, etc. This is why LEDs are traditionally driven with a bias current and not a voltage current; the manufacturer provides an average forward voltage and its dispersion with the nominal current. Nowadays, the nominal current for most of the components is 350 mA with a corresponding forward voltage of typically 3 to 3.5V; the maximum current can go up to 1, even 1.5 A. The component is either biased with a direct or a pulse-width modulation current. As addressed previously, most of the components have a protection against direct or inverse electrostatic discharges up to 10 or 20kV.

From the thermal point of view, the component is characterized by:

– the thermal resistance from the junction to the slug R_{j-s}; the typical data of current components ranges from 5 to 15 K/W, but multi-chip components now have a thermal resistance below 5 K/W;

– the forward voltage temperature coefficient in V/°C (a few mV/°C);

– the maximum junction temperature (150 or 180°C in general);

– the maximum temperature range of ambient air for the nominal operating mode.

As explained in section 5.3, this last characteristic is really dependent on the thermal resistance R_{j-s}, but also on the total thermal resistance R_{j-a}. Typically, the operating range is from -40/+135°C to -20/+85°C depending on applications. At high ambient temperatures, and depending on the total thermal resistance, it is no longer possible to use the nominal current due to overheating of the junction temperature component; the manufacturer generally recommends a maximum current as a function of ambient temperature for different values of total thermal resistance.

5.5.2. *Optical characteristics*

Those different characteristics are addressed in more detail in Chapters 6 and 7 for color issues and we will limit ourselves to packaging-related aspects.

A LED is characterized by the luminous output flux (expressed in lumens), which corresponds to the total emitted luminous power (expressed in Watts) convoluted with the eye perception spectrum between 380 and 780 nm. There is tough competition between manufacturers to deliver the highest luminous flux for a single chip per package. Today, a typical luminous flux for a nominal current is 100 to 200, even 250 lm at the maximum current driving for a single chip component. In general, the manufacturer also gives the luminous flux as a function of junction temperature (because the luminous flux depends on heat dissipation) and the curve of the luminous flux versus bias current (with the component's slug held at 25°C). This last curve is the most important because it takes into account the thermal resistance of the carrier separately from the motherboard support which is maintained at 25°C. In terms of efficiency, the LED's total wall plug efficiency in lumen/Watt corresponds to the luminous flux out of the driving electrical power; this yield is highly dependent on heat dissipation and thus on the bias current. Today, for a driving current of 350 mA, the best efficiencies reported by manufacturers at the lab scale are about 150 lm/W for a single chip [LED 08]. It seems to be difficult to compare the efficiencies of components because, in addition to thermal aspects, total efficiency depends on the driving current and on the chip's surface area. A relevant comparison would not be with the given driving current but with the equivalent current density (i.e. the total current with respect to the surface of the chip).

The manufacturer gives the emission spectrum for the visible range at the nominal operating current and with the slug maintained at 25°C. This spectrum depends on the primary emission peak and on the phosphor conversion material(s). For white LEDs, it is traditionally characterized by two values, which are the correlated color temperature (CCT) and the color rendering index (CRI) (see Chapter 7). To simplify matters, the CCT represents a color balance when the source is in direct sight and the CRI is the rendering of a reference color panel when these colors are illuminated by the light source. A CCT of about 6,000 K corresponds to a typical solar emission spectrum and is called cold because there is a blue dominance compared to red and green. A color temperature of 3,000 K is more balanced and is called warm because it provides a more pleasant lighting atmosphere. If the CCT is generally given, the CRI is not always specified because it is generally not to the advantage of the LED systems when compared to incandescent or fluorescent light bulbs.

We have already shown in section 5.2 that, depending on packaging processes, the deposition of the phosphor material is more or less conformal. A conformal phosphor deposition, in other words a constant thickness above the whole chip surface area, leads to a shorter path for emissions perpendicular to the LED surface than for off-axis emissions. It translates into a colder correlated color temperature in the LED axis compared to an observation in the off-axis direction. In the opposite case, for instance when the phosphor's thickness has a dome shape, the phosphor's secondary contribution to the emission spectrum is stronger in the beam perpendicular to the emission surface than in off-axis emission. This phenomenon creates a warmer color in direct beam than off-axis. Lastly, a final effect is sometimes observed when the phosphor deposition is not uniform with a very thin layer on the edges of the chip; a dominant blue color is observed at a very specific angle for reasons discussed above and is known as the "blue ring" effect. An accurate control and design of the phosphor material could thus allow the control of the uniformity of the colorimetric quality of the light emission. This remains a weakness in LED packaging.

Finally, the spatial radiation pattern of the LED's luminous flux, in other words light intensity distribution versus the observation angle, is usually given. Generally, it is provided in polar coordinates r(θ), where r is the maximum light flux in percentage and θ the observation angle. The most standard spatial radiation pattern is the Lambertian pattern; this is the situation where the visual perception does not depend on the observation angle of the planar extended emission source. The spatial emission diagram pattern can be tailored using lenses of a specific shape on each component.

5.5.3. *Binning*

Binning takes into account parameters such as the luminous flux (in lumen), the CCT (in Kelvins), and the forward voltage (in volts). The dispersion of these parameters for several components, deemed as identical from a fabrication process point of view, is huge. Today, manufacturing processes are not mature enough and as a matter of fact, each component is individually tested at the end of the manufacturing process and sorted in different classes/binnings. Historically, this is partly due to the car industry because very demanding tolerances on the specifications have been required of the electronic components. Considering the luminous flux, the eye is sensitive to a variation of 20 to 30%, whereas common dispersion right after fabrication can go up to a factor of 3. The situation is similar for CCT; the human eye can detect a variation of 1 to 2 nm in the main emission peak in the green domain when the common dispersion is typically about 15 nm. In industry, the classification is done by giving priority to some parameters; a first selection is for instance done in the color's temperature with a black body as a reference and defining the white: cold white (4,500 to 10,000 K), neutral white (3,500 K to 4,500 K), and warm white (2,650 to 3,500 K). In the same category, components are then sorted into sub-categories depending on their luminous flux and each sub-category gives a tolerance in forward voltage data. Thus, no components are rejected, but components with the best performances become the most expensive.

5.5.4. *Reliability*

Intrinsic aging of the chip aside, a LED undergoes a series of thermomechanical and environmental stresses that can lead to a reduction in the lifetime of the component. These stresses are present right from the beginning at the mounting step on the motherboard or in the system, and others appear throughout the rest of the LED lifetime. They can be identified and have different origins (non-exhaustive):

– electrical: electrostatic discharges, high current bias;

– thermal: slow temperature variations or thermal shock;

– mechanical: shock, vibrations, etc.;

– environmental: solvents, humidity, etc.;

– optical: UV, etc.

All these stresses can damage the component on their own but also through a synergy effect.

5.5.4.1. *Stresses during the integration*

As discussed throughout the chapter, the component can be attached to a motherboard, whether it is metal or polymer-based, and then integrated into a system. Like any electronic component, the most common mounting processes are gluing and soldering. It is virtually possible to use any common solder metal to attach the chip to the motherboard. Most used leadfree alloys are Sn-Ag or Sn-Ag-Cu-based alloys with a melting point of around 220°C. Implementation methods using solders usually involve deoxidizing fluxes during the soldering step and cleaning processes that can potentially damage the lens or encapsulant materials. For gluing, Ag-based adhesives are commonly used. An advantage of gluing is that implementation only needs a moderate temperature (100-150°C or even ambient temperature), which reduces thermal mismatch stresses; moreover, these glues or pastes are soft materials, which may also accommodate to some extent the thermal expansion mismatch stresses more easily. Let us recall that solders are favored over adhesives from the thermal point of view.

5.5.4.2. *Operating stresses*

These stresses obviously depend on the application and thus there are no existing lighting standards. Restrictions for the car industry are tougher than for backlighting, for instance. The most common tests are:

– thermal cycling: -20/+85°C or -20/+125°C (1,000 cycles);

– dry air temperature storage: 85 or 100°C (1,000 hours);

– humid air storage: 85°C/85% relative humidity;

– bias current aging: in temperature and/or humidity.

From a packaging point of view, it is important to remember that metal alloys used for soldering are chemically stable and have a good thermal conductivity and temperature strength. However, they usually require a complex implementation, they are rigid materials and do not well accommodate thermomechanical stresses. On the other hand, adhesives are polymer materials that are thermally and chemically not as stable (especially with humidity), but their implementation is simpler and can be done at low temperatures; finally, they are soft materials that enhance thermomechanical stress relaxation. Aside from electrical contacts, the most common failures encountered in LEDs are related to:

– humidity in polymers that produce an thermal expansion mismatch of polymer and metallic/inorganic materials leading to cracks and delaminations at the interfaces;

– temperature and CTE mismatch between constitutive materials (metals, polymers and ceramics) that also leads to cracks and delaminations.

This is why all built LEDs undergo qualification, reliability and accelerated aging tests depending on their specific applications. We have to note that most failures are thermally activated.

5.6. Conclusion and trends

Overall, the packaging evolution follows one of the electronic components: not only miniaturization needs but also the necessity to yield the lighting function of a chip in a lighting system with the simplest and cheapest technology and the best possible efficiency. This evolution is related to the chip's evolution, its internal design (lateral/vertical configuration), and more and more demanding performances in terms of electrical, thermal and optical power generated.

Pin components evolution towards surface mountable "leadform" and "leadless" components become standards for reasons of compactness. We can recall that, as for 2D association in common microelectronics, future trends will tend towards chip scale packaging where the component's size is reduced and its surface approaches the chip's surface. This will lead to collective packaging at the wafer level called the "wafer level package" to produce single components in cost-efficient processes. "System in package" processes like "chip-on-board" will integrate several chips or a chip matrix in a single component or directly attached to a motherboard, in addition to passive and/or active components like "drivers", etc. Unlike common electronics, the evolution towards 3D packaging seems more complex since light extraction through one or more sides of the LED becomes an issue in 3D packaging.

Like all power electronics components, reliability and efficiency made it necessary to take into account heat dissipation not only at the motherboard level, but also directly in the component itself. As a matter of fact, the output optical power extracted from a component is directly limited by heat dissipation. Depending on the packaging processes, the use of insulating or conductive materials with high thermal conductivity like alumina and copper is common. It is likely that in the near future, better performing heat dissipation materials like AlN, carbon fiber or diamond composites, etc., which are currently too expensive, will be widely used, since cheap collective manufacturing processes and mass production will be adopted. Wall plug efficiency for white LEDs, being nowadays around 20-25%, will certainly be improved and will reach 50% in the next decade. Thus, heat dissipation is and will still be a challenging issue with a necessary cost/performance trade-off.

Finally, concerning light extraction issues, this being the main functionality of a LED, we showed that primary optics like encapsulants and lenses tend to use more temperature-stable, UV-resistant materials and slowly abandon epoxies for silicones. Moreover, these materials can easily be shaped depending on the desired spatial radiation pattern using common injection-molding processes. Silicones have a moderate refractive index and it is likely that new formulations will appear with a silicone matrix and inorganic fillers having a high refractive index to improve light extraction. In addition to a high refractive index, the filler should have a nanometric size much smaller than the considered wavelength in order to limit absorption and diffraction phenomena. Concerning phosphor implementation, the trend is to deposit and implement it on the chip before the dicing of chip from the wafer or, on the other hand, to leave it away on the primary optic where the temperature is lower and the conversion efficiency is higher. Furthermore, for aging issues discussed above, another trend is to reduce or even remove the organic matrix to implement only the active inorganic part of the phosphor.

Concerning level 2 and 3 packaging for lighting systems, the main advantages of LED systems compared to other sources like incandescent or fluorescent lightbulbs are numerous. The compactness of LEDs offers many design and packaging solutions to create an infinite number of 2D or even 3D architectures for lighting systems. The low voltage driving bias for these systems offers a large number of opportunities for fast dynamic driving allowing an extended control of the lighting through adapted electronics or even sensor control. Finally, LED manufacturing technologies are very close to silicon semiconductor fabrication, and could be easily associated with all microelectronic Si components, whether they are active control systems, logical circuits or even MEMS or MOEMS sensors.

5.7. Appendix

5.7.1. *Physical properties of materials*

The following table presents a few of the most used materials in packaging and their physical properties:

– Values given are not absolute and they depend on the purity of the manufacturing process. The best and most accurate references remain technical data sheets given by manufacturers.

– Regarding ceramics, the values are given for insulating materials (except GaN). Most of the nitride and carbide ceramics are semiconductors and can be doped to become conductive materials.

Materials	T_f (°C)	α (ppmK^{-1})	ρ ($\mu\Omega$.cm)	K_T (W/m,K)
Pure elements				
Cr	1,860	6.5	13,2	91
Mo	2,615	5.1	5,7	137
Ta	2,980	6.5	13,5	58
Ti	1,667	8.9	54	22
W	3,387	4.5	5,4	174
Zr	1,852	5.9	44	23
Ag	960	19.5	1,6	420
Al	660	22	2,6	200
Au	1,063	14	2,4	300
Cu	1,083	17	1,7	400
Ge	937	5.3	semiconductor	60
In	157	33	8,4	80
Mo	2,610	5	5,3	140
Ni	1,453	14	6,9	90
Pb	327	29	22	35
Pd	1,552	11	10.8	71
Pt	1,769	9	10.6	73
Si	1,414	2.6	semiconductor	80 to 150
Sn	232	20	12	64
Zn	419	30	5.7	113
Solders				
Sn-Pb 60/40	183	26	14	34
SnAg 96/4	221	11	30	33
Au-Sn 80/20	280	15	16	58
Au-Ge 88/12	356	12		44
Au-Si 97/3	363	11.5		216
In-Sn 50/50	120	20	14.5	34
Alloys				
Kovar (Fe-Ni-Co)	>1,000	5	50	16

Ferronickel FN42	>1,000	2.6		16
Nichrome	1,395	13	100	13
Cu-Al (E)	550	23		180
Ceramics				
AlN	2,300	2.1 to 4.4	Insulating material	100 to 230
Alumina	2,050	6.5	Insulating material	20 to 40
BeO	2,530	6.4 to 8.3	Insulating material	300 to 270
CBN		3.7	Insulating material	600
Diamond		1 to 4	Insulating material	900 to 2,000
HBN Adv Ceram		4 to 6	Insulating material	33 to 59
Si_3N_4	1,750	3.3	Insulating material	170
SiC	2,200	3.8	Insulating material	60-270
SiO_2 (quartz)	1,710	0.5	Insulating material	1,4
GaN	> 2,500	3.2	semiconductor	130
sapphire	2,050	6.5	Insulating material	40
Composites				
Al-SiC	570	6.5 to 7.3	4	160 to 180
Cu-Mo	1,083	7	/	170
Cu-W 20/80	1,083	7	2.5	250
Cu-Mo-Cu (1,6,1)	1,083	6.4	/	230
Conductive adhesives or pastes				
Epoxy-Ag	/			1 to 30
Glass-Ag	> 400			50 to 100
Silicone-Ag	/			~ 1 to 4
Polyimide-Ag	/			~ 2

5.8. Bibliography

[AMA 07] http://ecx.images-amazon.com/images/I/51IX-9RGWrL._SS400_.jpg.

[ARI 02] ARIK M., PETROSKI J., WEAVERY S., "Thermal challenges in the future generation of solid state lighting applications: light emitting diodes", *The Eighth Intersociety Conference on Thermal and Thermomechanical Phenomena in Electronic Systems*, ITHERM 2002, p. 113-120.

[CUR 07] http://www.curamik.com.

[EPO 07] http://www.polytec.fr/EPOTEK-1.

[GLO 07] http://www.globec.co.uk.

[HAR 04] HARPER C.A., *Electronic Packaging and Interconnection Handbook*, 3rd Edition, McGraw-Hill, New York, Chapter 3, 2004.

[HOL 62] HOLONYAK N. Jr. and BEVACQUA S. F., "Coherent (visible) light emission from GaAs P junctions", *Applied Physics Letters*, vol. 1, p. 82–83, 1962.

[HYM 07] http://www.hymite.com.

[JAC 93] JACOBSON D.M., *Principles of Soldering & Brazing*, ASM International, 1993.

[KRA 07] KRAMES M.R., SHCHEKIN O. B., MUELLER-MACH R., MUELLER G. O., ZHOU L., HARBERS G., CRAFORD M.G., "Status and future of high-power light-emitting diodes for solid-state lighting", *Journal of Display Technology*, vol. 3, no.2, June, p. 160-175, 2007.

[KRA 99] KRAMES M.R. *et al.*, "High-power truncated-inverted-pyramid (AlxGa1-x)0.5In0.5P/GaP light-emitting diodes exhibiting >50% external quantum efficiency", *Applied Physics Letters*, 75[16], p. 2365-67, 1999.

[LED 08] http://www.ledsmagazine.com/features/5/10/5,
http://www.ledsmagazine.com/news/ 5/11/20.

[LIC 05] LICARI J.J., SWANSON, D.W., *Adhesive Technology for Electronic Applications: Materials, Process, Reliability*, William Andrew Publishing, Chapter 3, p. 95-168, 2005.

[LIU 93] LIU J., *Conductive Adhesive for Electronics Packaging*, Electrochemical Publications Ltd., Port Erin, Isle of Man, UK, 1999.

[LUM 07] LUMILEDS, *Understanding Power LED Lifetime Analysis*, White Paper, http://www.lumileds.com/docs/docs.cfm?docType=16, 05/22/2007.

[MON 07] http://www.led-fr.net/caracteristiques_led.htm.

[MUR 07] MURPHY T. *et al.*, “A silicon wafer packaging solution for HB-LEDs”, *Workshop on Manufacturing LEDs for Lighting and Displays*, Berlin, September 10-11, 2007, EPIC/SPIE Europe.

[NUE 69] NUESE C.J., TIETJEN J.J., GANNON J.J., and GOSSENBERGER H.F., “Optimization of electroluminescent efficiencies for vapor-grown GaAs.sub.1-X P.sub.X diodes”, *Journal of Electrochemical Society, Solid State Science*, vol. 116, no.2, p. 248, 1969.

[PET 01] PETIT J.P., “Dissipation thermique dans les systèmes électroniques”, *Techniques de l’ingénieur*, vol. E 3952, February 2001.

[SAI 05] SAINT MARTIN X., “Packaging des circuits intégrés”, *Techniques de l’ingénieur*, vol. E 3400, February 2005.

[SCH 06] SCHUBERT E.F., *Light Emitting Diodes,* Second edition, Cambridge University Press, Chapters 5 and 9, 2006.

[SCH 04] XI Y., SCHUBERT E.F, “Junction–temperature measurement in GaN ultraviolet light-emitting diodes using the diode forward voltage method”, *Applied Physics Letters*, Vol. 85, no. 12, 20 September 2004, pp. 2163-2165.

[STEI 02] STESTEIGERWALD D.A., BHAT J.C., COLLINS D., FLETCHER R. M., HOLCOMB M. OCHIAI, LUDOWISE M.J., MARTIN P.S., RUDAZ S.L., “Illumination with solid state lighting technology”, *IEEE Journal on Selected Topics in Quantum Electronics*, vol. 8, no.2, March/April 2002.

[ZEH 07] ZEHNDER U., “GaInN LEDs, Straight way towards solid state lighting, manufacturing LEDs for lighting and displays”, *Proceedings of SPIE*, vol. 6797, Thomas P. Pearsall, 26 September 2007.

[ZWE 06] ZWEBEN C., “Thermal materials solve power electronics challenges”, *Power Electronics Technology*, February 2006, pp. 40-47.

Chapter 6

Photoelectric Characterization of Electroluminescent Photodiodes

6.1. Photometry of LEDs

No science without measurement! The metrological characterization of LEDs is thus an uncircumventing step in every activity of research and development implying these components. There is also no serious production activity without quality control, which justifies once again to going through this step in LED manufacturing. Finally, at the commercial level, a comparison between products is only possible if these are correctly and uniformly characterized for every product. In section 6.1, we will focus our interest on the photometric characterization of LEDs i.e. on their luminous behavior (the term "photometry" being here taken in its larger sense, in opposition to its more restrictive sense defined hereafter). Section 6.2 will be devoted to their electrical and thermal characterization. Only the individual inorganic LEDs will be considered, to the exclusion of clusters or networks of LEDs or also LEDs having a large emission surface as organic LEDs (OLEDs). The works of the CIE (Commission Internationale de l'Eclairage – International Commission on Illumination) will be our main guide.

Chapter written by Christian EUGÈNE and Jean-Michel DESWERT.

Like every light source, an LED diode emits a radiation carrying energy. This must be characterized globally (*radiometry*) or spectrally, i.e. as a function of the wavelength λ (*spectroradiometry*). If the radiation is confined to the visible range and above all if it is dedicated to a lighting application, for the global parameters, we will prefer to use *photometric units*. If, furthermore, we are particularly interested in colored aspects, we will use concepts relevant to *colorimetry* (this question is more specifically addressed in Chapter 7). Finally, the *spatial distribution* of the radiation is another important aspect of the characterization of a source.

6.1.1. *Recap of fundamental knowledge*

6.1.1.1. *Link between radiometry and photometry*

The photometric and colorimetric quantities of a radiation are linked to perception. They are deduced from its energetic spectrum by non-ambiguous mathematical relations. For instance, to go from an energetic quantity G_e to a photometric quantity G (the subscript "v" for "visual" should be used, but it is often omitted by default; Conversely, for an energetic (or radiometric) quantity, the subscript "e" is more imperative), we have, for each wavelength λ, to weight the corresponding energetic spectral density of the radiation $dG_e/d\lambda$ by the relative spectral sensitivity $V(\lambda)$ of the "photometric standard observer" of the CIE (we also call this quantity the "relative spectral luminous efficiency function in photopic vision" [CIE 05] (see Figure 6.1)), and to integrate the whole as a function of λ. We will still multiply this result by a coefficient K_M which is 683 lm/W (lumens per watt) in the international system of units – SI. In SI units (the integration limits are this of the visible region as it ensues naturally from $V(\lambda)$ – they are 380-760 nm if $V(\lambda)$ is rounded to four decimals), we have:

$$G = 683 \int_{visible} V(\lambda) \frac{dG_e}{d\lambda} \, d\lambda \qquad [6.1]$$

In lighting, the photometric quantities will be, as said before, preferred with regard to radiometric (or energetic) quantities. Of course, for applications other than in the visible range (we think for instance of infrared LEDs used in communications), these radiometric quantities are exclusively in use.

To know the definitions of the different radiometric, photometric and colorimetric quantities, we will consult the International Lighting Vocabulary of the CIE [CIE 87a]. The basic concepts in colorimetry are described in the CIE 15 publication [CIE 04]. Unless otherwise mentioned, we will limit ourselves in the following to the photometric quantities. It will be easy to transpose, *mutatis mutandis*, to energetic quantities.

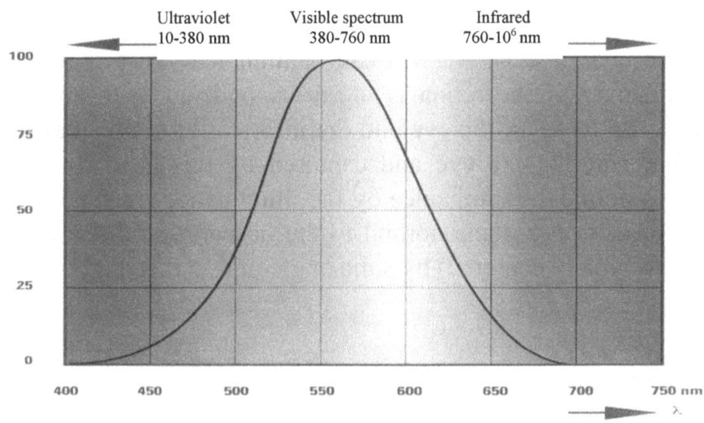

Figure 6.1. *Relative spectral sensitivity V(λ)*

6.1.1.2. *The importance of luminance*

Lighting has no sense without referring to the perception that man has of his visual environment. How can we characterize this perceived visual environment? In each point of the space where the observer stays and in each observing direction, the retina receives and routes to the brain a message in relation with the radiation coming from this direction. The visual information contained in this message has a nature both quantitative (ordinary expressed by words as luminosity, brightness, brilliance, etc.) and qualitative (color or hue). These two aspects are addressed by two distinct metrics: photometry for the quantitative (this word is thus here used in its restrictive sense, the foundation of which is given by relation [6.1]) and colorimetry for the qualitative. Three parameters are actually sufficient for characterizing this information because for day or photopic vision (concerned in the perception of light sources), the cells of the retina, the cones, are differentiated by three different pigmentations: the vision is trivariant. Evidently, if the quantitative aspect is characterized by one value only (the luminance considered below), two parameters therefore remain for the qualitative aspect (the chromatic coordinates). These two metrics are the tools used for characterizing the visual perception and therefore for objectively formulating the photometric (this word is here to be taken in its large sense) characteristics of every light source.

Let us return to the quantitative message taken in charge by the photometry and let us refine our interpretation in order to deduce from it a representative quantity that we will call "luminance". Just as for a camera, each photoreceiver of the retina is in correspondence with an elementary sighting cone centered on the eye and the information it receives represents the light flux entering the eye and captured by this cone. It is therefore appropriate to define the luminance by the illuminance received at the point of observation and in a plane normal to the observation direction (because the sight is essentially centered) by solid angle unit in this direction:

$$L = \frac{dE_{\perp}}{d\omega_{received}} \qquad [6.2]$$

Let us recall that the illuminance E in a point of a receiving surface is by definition the flux received by surface unit in this point. It is expressed in lux (lx), in photometric units (1 lx = 1 lm/m^2) or in W/m^2 in energetic units (the term is here irradiance):

$$E = \frac{d\Phi}{dS_{received}} \qquad [6.3]$$

It is easy to show that, through the play of the eye optic, L is proportional to the retinal illuminance, which ends the justification that luminance is the fundamental parameter of the quantitative perception. Relation [6.2] is usually called the "second" definition of the luminance because it is formulated from the observer. Its "first" definition implies the notion of intensity I emitted by a defined "point" of a source of a certain extension, i.e. the elementary emitting surface dS_{emit} around this point. This intensity I is, by definition, the flux emitted by this point by a solid angle unit in a specified direction. This quantity is thus vectorial:

$$I = \frac{d\Phi}{d\omega_{emit}} \qquad [6.4]$$

I is expressed in candela (cd) (1 cd = 1 lm/sr) in photometric units or in watts per steradian (W/sr) in energetic units.

The luminance L of the considered point of the source is thus its intensity by surface unit normally projected on the considered direction:

$$L = \frac{dI}{dS_{\perp emit}} \qquad [6.5]$$

It is easy to verify the equivalence of both definitions [6.2] and [6.5]. The latter actually expresses the fact that luminance is a notion attached to a source, but it requires its localization in the space, which is not needed as expression [6.2] shows (only the direction is indicated without indication of the distance between the source and the observer), and is not even always possible (e.g. sky, fog).

The luminance is expressed in candela per meter square (cd/m^2) in photometric units, as it results from expression [6.5]. In radiometric units, the term is radiance and its unit is $W/sr\text{-}m^2$.

6.1.1.3. *Intensity and flux*

We showed above that it is the luminance of the "points" of the visual surrounding environment which fundamentally characterizes our quantitative perception of this environment. Multiplying the luminance of each of these "points" by their elementary (projected) surface, we can also speak about their intensity in the direction of observation. It is the parameter we privilege in characterizing a light source. In order for this source to be reduced to one "point", in which case one value only for the intensity characterizes it for each direction, it is necessary that this source be small in comparison to the observation distance (intensity is an "at infinite" concept). Otherwise, this unique value can only be an average value, still depending on this distance.

It is here appropriate to mention a remarkable geometric law largely used in lighting which links the illuminance produced by a punctual source in a point of a receiving surface to the intensity of this source in the direction of this point:

$$E = \frac{I \cos\theta}{d^2} \qquad [6.6]$$

where d is the distance and θ the angle of incidence.

The manner in which the intensity varies as a function of the direction is an essential characteristic of every source (assumed to be punctual as explained), in particular of LEDs. We speak here about the radiation diagram. The most usual representation is a polar diagram in the 3D space resulting from the tracing in each direction of a vector ray exiting the source, the amplitude of which is proportional to the intensity in this direction. Most often, a reference axis is defined (e.g. axis of rotation or of symmetry) and a 2D curve is given for a series of meridians passing through this axis.

The volume delimited by this polar diagram represents the total flux Φ emitted by the source (in lumens (lm), in photometric units or in watts (W) in radiometric units). More exactly (this formula being of application only for a punctual source):

$$\Phi = \int_0^{4\pi} I d\omega \qquad [6.7]$$

In the general case of one or several source(s) of any extension, the following relation is appropriate:

$$\Phi = \int_A E dA \qquad [6.8]$$

where the illuminance E is integrated on the whole surface A of any closed volume surrounding the source(s).

6.1.2. *Parameters of interest*

The considerations above evoke most of the parameters of interest characterizing LED diodes. Let us give a synthesis of these.

The first quantity is the total flux Φ emitted by the source. The manner in which this flux is distributed into the 4π steradian of the space is characterized by its radiation diagram in intensity I (also called intensity spatial distribution or radiation indicatrix), which assumes an observation at large distance. If we divide this intensity by the apparent surface of the source in the considered direction, we obtain the luminance in this direction (more exactly, the average luminance if this surface is not homogenous). We can also be interested in the partial flux in a given angular sector. Φ and I are generally expressed in photometric units (lm and cd).

Now that the source is quantitatively defined, the qualitative aspect of the radiation is completely defined by its spectral content (relative to its maximum), i.e. the relative spectral density (or distribution) curve of any radiometric quantity of this radiation: its derivative as a function of λ, normalized to its maximum. In a more synthetic way, it is sufficient to determine the two chromatic parameters (in any of the numerous possible formulations) of the radiation which are relevant to colorimetry.

We will examine all this more closely in the following text from a metrological point of view. We will successively address the measurement of the intensity and the related spatial distribution, of the flux and finally of the spectral content. First of all, we will specify the requested qualities of the measuring instruments. Our main reference will be the CIE 127 publication [CIE 07a].

6.1.3. *Required properties of photometers/radiometers*

In accordance with our interest in either photometric or radiometric quantities, the basic instrument will be a photometer or a radiometer, which respectively measures the illuminance (in lx – it is thus a lux meter) and the irradiance (in W/m^2). These instruments are composed of a measuring head comprising a detector, a filter and an entrance opening, completed by an electronic circuit to amplify and measure the electric signal at the detector output.

6.1.3.1. *Detectors*

The most often used detector is a silicon photodiode, the short-circuit current of which is the useful output signal called a photocurrent. Its spectral sensitivity covers a bandwidth from the near-ultraviolet to the near-infrared, crossing the whole visible range with a peak value around 900 nm and an abrupt fall around 1,100 nm. It shows a good linearity in amplitude (the photocurrent is proportional to the entrance flux over several decades) and a good thermal stability in the visible region.

6.1.3.2. *Angular and spatial response*

A photometer/radiometer must ideally respect the "cosines law" [6.6], i.e. that its response for a defined parallel incident beam must depend on the cosinus of the angle of incidence with regard to the normal to the reception surface. This requirement is particularly significant for applications where the incident light covers a large angular field, as is the case, for example, for the measurements of the integrated sphere method (see section 6.1.5.1.2). The manufacturers propose in this way "cosinus corrector" devices, removable or otherwise. In intensity measurements, this problem is less critical because the incident light covers a narrow angular field, near the normal.

For the measurement of "averaged LED intensity" (see below) where we work in the near field, it is still necessary that the active surface of the detector has a good spatial uniformity in order that all the radiations entering the opening are measured with the same weight.

6.1.3.3. *Spectral response*

Let s(λ) be the spectral sensitivity of the detector (filter included), i.e. the ratio, at the wavelength λ, between the output electric signal Y (current or voltage according to the case) and the input luminous quantity X_e expressed in energetic units (usually radiant power, i.e. the flux or its spatial density): $s(\lambda) = dY(\lambda)/dX_e(\lambda)$. If the incident radiation presents a spectral distribution $X_{e\lambda} = dX_e/d\lambda$, the output electric quantity Y thus has for an expression:

$$Y = \int_0^\infty X_{e\lambda}\, s(\lambda) d\lambda \qquad [6.9]$$

In the case of a radiometer (Y representing a radiometric quantity), s(λ) must be as constant as possible on the spectral waveband of the incident radiation.

On the contrary, as regards a photometer (Y representing a photometric quantity), s(λ) must be, in relative value to its maximum ($(s(\lambda)_{rel})$), as close as possible to V(λ). We speak here of "spectral matching". The error obtained in the case of mismatch depends on the spectral deviation between the source under test and the source against which the instrument was calibrated. This error would be zero if both sources had the same spectrum. This question is studied in [CIE 82] and [CIE 87b]. Other procedure recommendations for spectral characterization of detectors are given in [CIE 84a].

6.1.3.3.1. Mismatch correction by the four integrals method

The development hereafter presents the rigorous manner of applying a corrective factor to the measurement of illuminance, taking the detector mismatch into account. The requirements are multiple: the spectral sensitivity of the detector, the spectral content of both the source under test and the lamp against which the instrument is calibrated.

Let us take a lux meter of spectral sensitivity s(λ), giving the measure m_T of an illuminance E_T to be tested and the measure m_R of a reference illuminance E_R (standard). The energetic spectral densities of T and R are assumed to be known. The following results are obtained:

$$\frac{E_T}{E_R} = \frac{m_T}{m_R} * \frac{\int E_{e\lambda T} V(\lambda)d\lambda}{\int E_{e\lambda T} s(\lambda)d\lambda} * \frac{\int E_{e\lambda R} s(\lambda)d\lambda}{\int E_{e\lambda R} V(\lambda)d\lambda} \qquad [6.10]$$

The factor containing the four integrals mixes the spectral densities of E_T and E_R with s(λ) and V(λ). It is the searched corrective factor (hereafter called F) since if F=1, the ratio between the illuminances equals the ratio between the measurements. This formula results from:

– the basic photometric law:

$$E = K_M \int E_{e\lambda} V(\lambda)d\lambda \qquad [6.11]$$

– the transfer function of the lux meter:

$$m = \int E_{e\lambda} s(\lambda) d\lambda \qquad [6.12]$$

both applied to T and R.

In the corrective factor, all the spectral densities and sensitivities can be expressed equally in absolute or relative values because they appear simultaneously on the numerator and on the denominator.

The calibration under reference conditions allows the constant c of the lux meter to be known:

$$c = E_R / m_R \qquad [6.13]$$

and thus:

$$E_T = c\, m_T\, F \qquad [6.14]$$

The corrective factor F is specific for each spectral distribution to be tested. It is unitary only if the matching is perfect (s(λ) proportional to V(λ)) or if the lamp under test has the same relative spectrum as the reference lamp. Applied to LEDs, this comment leads to the idea of collecting a series of standard LEDs centered on the main available colors in order to enable the direct comparison of the LED under test to a reference LED of similar spectrum and, in such a way, to bypass the need to determine the corrective factor. Currently, these reference LEDs are still rare and expensive, but it is foreseeable that their availability will increase over time.

The long and tedious correction by the method of the four integrals, which requires multiple spectral determinations, suggests a direct determination of the searched photometric quantity by application of formula [6.1] relating this quantity to the corresponding spectral energetic density, which is determined by spectro-radiometry in accordance with one of the methods shown in section 6.1.6.

6.1.3.3.2. f_1' number of a lux meter [CIE 87b]

The spectral mismatch of the photometer is characterized in a synthetic manner by the f_1' number. This index expresses in a particular way the difference of the measurements obtained in the case of a source specified hereafter, depending on whether we use the concerned photometer or a perfectly matched photometer. The specified source is the illuminant A of the CIE (tungsten filament lamp with a color temperature of 2,856 K – the color temperature of a lamp is the temperature of a black body of the same relative spectrum) which is commonly used for the calibration of photometers. This index, which relates $s(\lambda)_{rel}$ to $V(\lambda)$ for the spectral distribution $S_{\lambda,A}$ of the illuminant A, is defined as follows:

$$f_1' = \frac{\int \left| s^*(\lambda)_{rel} - V(\lambda) \right| d\lambda}{\int V(\lambda) d\lambda}$$

with:

$$s^*(\lambda)_{rel} = s(\lambda)_{rel} \frac{\int S_{\lambda,A} V(\lambda) d(\lambda)}{\int S_{\lambda,A}\, s(\lambda)_{rel}\, d(\lambda)} \qquad [6.15]$$

We observe that this f_1' number, which is always positive, is zero in the case of perfect matching. As its definition doesn't imply the spectral distribution of the radiation under test, it cannot be an error or corrective term to be applied. It has nevertheless been shown that this number is representative of the limits of the measurement error we could make for a source of "white" light, on the condition that the photometer has been calibrated under this illuminant A. In the case of white LEDs, it is recommended that the instrument does not exceed $f_1' = 3\%$.

6.1.3.3.3. Measurement of colored LEDs

In the case of a mono-color LED (thus of narrow bandwidth), the error due to the detector mismatch can be high, even for a low f_1'number, because we take no more profit from the compensation effect of the errors over large spectra. The case is particularly acute for red or blue LEDs which are spectrally situated on the right and left flanks of the $V(\lambda)$ curve, far from the maximum (see Figure 6.2, where $V(\lambda)$ and $s(\lambda)_{rel}$ curves, with their maximum of similar wavelength by will of the detector manufacturer, are close together near their center but not necessarily on their flanks).

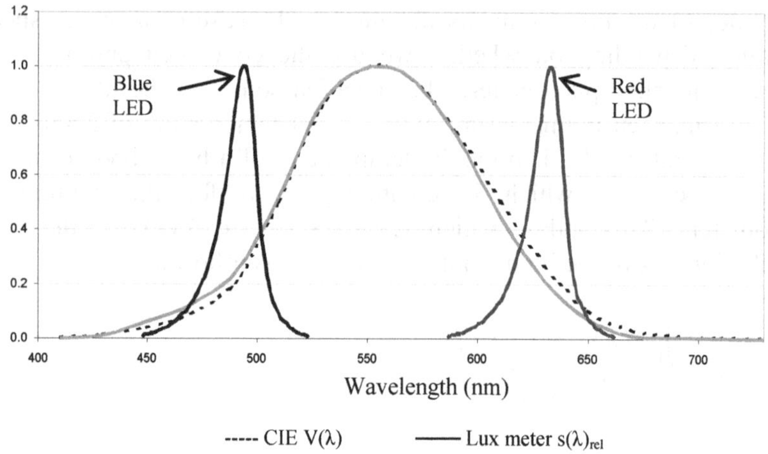

Figure 6.2. *Spectra of a blue LED and of a red LED superimposed on $V(\lambda)$ and $s(\lambda)_{rel}$*

To avoid the long and tedious calculation of the rigorous correction explained above, we will favor, as already mentioned, the direct application of formula [6.1] to the data obtained by spectroradiometry.

6.1.4. *Measurement of luminous intensity*

6.1.4.1. *Measurement in the far field*

As already said, the concept of intensity assumes a punctual source. This quantity is determined by a lux meter placed at a sufficient distance and the [6.6] law is applied, which contains the square of the distance and the cosinus of the tilt angle (it is usual to take a normal incidence in order to have a unit value for this cosinus).

No source is strictly punctual but becomes so if the distance between source and detector is large with regard to the largest dimension of the source. The practical recommendation is to take at least a ratio of 10 between these two quantities, or still more in the presence of very collimating optics in the front of the source, as is often the case with LEDs.

To achieve a normal incidence in each point of the detector surface, it is still necessary that the largest dimension of this is small with regard to the distance. This double condition of smallness both of the source and the detector with regard to the distance is usually called the "far field condition". If at least one of these conditions is not satisfied, we speak of "near field", for which the [6.6] law above is no longer valid.

It is not always easy to determine the distance between the source and the detector, on the one hand because the active surface of the detector cannot be accurately localized due to the presence of the matching filter, and on the other hand because of the effect of a lens often placed in front of the LED. In this latter case, we are reduced to assuming an arbitrary localization of the optical center in a specified point of the LED.

6.1.4.2. *Averaged intensity of an LED*

In the measurements of LEDs, it is frequent, in industrial practice where compactness of devices and manipulation easiness are favored, to be in near field conditions for which the quantity deduced from the square of the distance applied to the illuminance no longer represents the true intensity in the considered direction. It will actually be a conventional intensity called averaged intensity, the term "averaged" coming from the fact that all of the elementary emitting surfaces of the source (which is not more punctual) are concerned.

As the measurement result in near field is essentially dependent on the chosen distance, it is necessary to accompany this notion of averaged intensity with a precise description of the operating geometry. The CIE proposes two standard geometries. There are noted respectively A and B and the corresponding intensity is I_{LEDAv} and I_{LEDAe}, or I_{LEDBv} and I_{LEDBe} according to whether the photometric or the energetic intensities are addressed. Both standard conditions imply a detector with a circular input opening of 100 mm^2 area facing the LED and with its center aligned along the mechanical axis of the LED. The difference between both situations lies in the distance d between the LED and the detector (in both cases, the distance is measured from the LED front end to the detector input opening plane):

– condition A: $d = 316$ mm (thus $d^2 = 0.1\ m^2$),

– condition B: $d = 100$ mm (thus $d^2 = 0.01\ m^2$).

We thus obtain:

– in condition A: I_{LEDAv} (cd) = 0.1 E_v (lx),

– in condition B: I_{LEDBv} (cd) = 0.01 E_v (lx).

6.1.4.3. *Measurement of the intensity spatial distribution*

In order to determine the radiation diagram of the source, use is made of a two-axes goniometer able to explore every spatial direction. The lamp is situated at the center and the detector, placed at a sufficient distance in order to validly apply the law of the square of the distance, explores a series of meridians under different azimuths.

Methods of angular exploration (Figure 6.3)

a) To maintain a fixed attitude of the source and to describe half a circle (e.g. in the vertical plane) with the photodetector attached at the bottom of an arm rotating around the center of the source. In such a way, a meridian plane is explored which can be changed by pivoting either the source or the rotating plane around the vertical axis. The free space required for the meridian exploration will be as vast as the distance source-detector is large.

b) The source is placed at the rotation center of a double-orthogonal-axes mount and the detector is fixed. We thus economize on room, but this does not suit sources for which the flux depends on their position with regard to the vertical. This problem is not an issue with LEDs.

There is still a more elaborated method based on a rotating mirror, but it is not justified for LEDs.

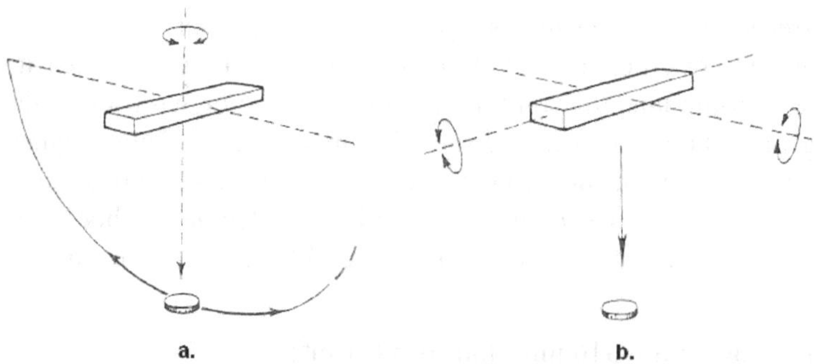

a.　　　　　　　　b.

Figure 6.3. *Methods of angular exploration*

6.1.5. *Measurement of luminous flux*

The usual quantity of interest is the total flux emitted by the source in all directions. In certain cases, we are also interested in the partial flux emitted in a given solid angle (for a punctual source), therefore ignoring the flux irradiated in undesirable directions, such as the back of the LED.

6.1.5.1. *Measuring methods for the total flux ([CIE 89])*

6.1.5.1.1. Goniophotometric method

The fundamental method for measuring the total flux Φ is to integer the intensity I in the 4π steradians of the space according to definition relation [6.7]. I is measured by a goniophotometric method described above (two-axes goniometer exploring azimuth angle φ and elevation angle θ). As already said, this definition assumes a punctual source (and thus a measurement at large distance).

This constraint is removed if definition [6.8] is used integrating the illuminance on the total surface of any volume surrounding the source. The measurement method explores, in a practical way, the surface of a sphere of any radius R (the lower limit being of course the requirement that the sphere contains the source, but also that the correct "cosinus" behavior of the detector is not imperilled by incidences that are too tilted). It is not necessary that the source be placed at the center of the sphere even if this is its usual position. As, in every case, the measurement comes to determine the illuminances on a sphere of radius R, the following relation is applied:

$$\Phi = R^2 \int_{\varphi=0}^{2\pi} \int_{\theta=0}^{\pi} E(\theta,\varphi) \sin\theta \, d\theta \, d\varphi \qquad [6.16]$$

The following transformation sequence can be useful in this calculation (see Figure 6.4 expressed for a unit radius):

$$\Phi = R^2 \int_{\theta=0}^{\pi} [\int_{\varphi=0}^{2\pi} E(\theta,\varphi) d\varphi] \sin\theta \, d\theta = 2\pi R^2 \int_{\theta=0}^{\pi} E(\theta)_{aver.on\,\varphi} \sin\theta \, d\theta$$

$$= 2\pi R^2 \int_0^2 E(x)_{aver.on\,\varphi} dx \qquad with \quad x = 1-\cos\theta \qquad (0 \le x \le 2) \qquad [6.17]$$

The calculation of Φ thus comes to measure the area under the curve expressing $E_{\text{aver.on }\varphi}$ as a function of x on a Cartesian diagram and to multiply this area by $2\pi R^2$.

The measuring angular intervals must be adequately chosen as a function of how acute the beam distribution is.

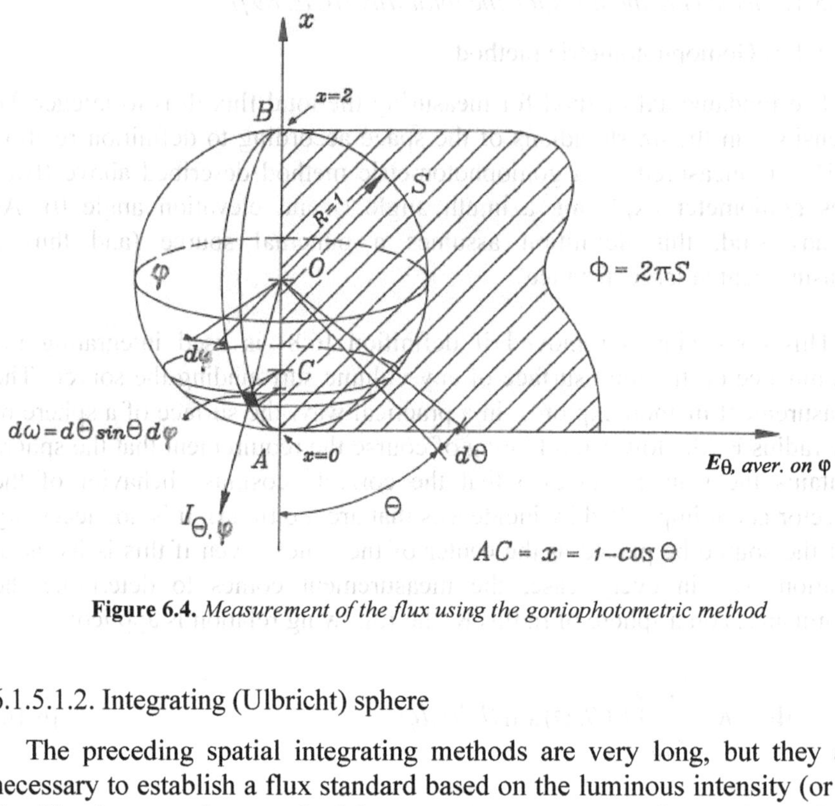

Figure 6.4. *Measurement of the flux using the goniophotometric method*

6.1.5.1.2. Integrating (Ulbricht) sphere

The preceding spatial integrating methods are very long, but they are necessary to establish a flux standard based on the luminous intensity (or on the illuminance of a standard lux meter). In contrast, the method of the integrating Ulbricht sphere makes this integration in an optical way in order to directly give the total flux of a source under test. However, it is a *comparison method* which assumes the availability of another source, if possible of similar spatial and spectral distributions, of known flux (standard lamp). If the sphere is often considered as a bulky piece of equipment when it has to contain lamps and luminaries of any dimensions, this objection is not an issue for LEDs.

Let us take a sphere of radius R (Figure 6.5) with the inner wall covered with a white, spectrally non-selective painting. The reflectance ρ (i.e. the ratio between the flux reflected and the flux incident on the point considered) of this painting is assumed to be high and constant in every point. This painting must also be perfectly matt, i.e. of ideal diffuse reflectance reflecting the light under constant luminance in all directions (we speak here about isotropic or uniform or Lambertian reflectance). The lamp (source S) for which the flux Φ is to be measured is placed in any location of the interior volume of the sphere.

Let A be an arbitrary point of the wall: the illuminance E on this point is due to the direct illuminance E_d of the source increased by the indirect reflection illuminance E_r due to the successive reflections on the inner wall of the sphere. We will show that, theoretically, the indirect reflection illuminance on the wall is independent of the position of point A and is expressed by:

$$E_r = \frac{\Phi}{4\pi R^2}\frac{\rho}{1-\rho} \qquad [6.18]$$

It immediately appears that illuminance E_r is proportional to the flux Φ according to a very simple relation. In order to prove it, let us start from the following equation giving the energetic balance:

Flux Φ emitted by the source = total flux absorbed by the sphere (of area S):

$$\Phi = (1-\rho)\int_S E\,dS = (1-\rho)\int_S E_d\ dS + (1-\rho)\int_S E_r\ dS \qquad [6.19]$$

However, Φ is still written according to its general definition [6.7]:

$$\Phi = \int_S E_d\ dS \qquad [6.20]$$

Therefore, the balance becomes:

$$\int_S E_r\ dS = \Phi\frac{\rho}{1-\rho} \qquad [6.21]$$

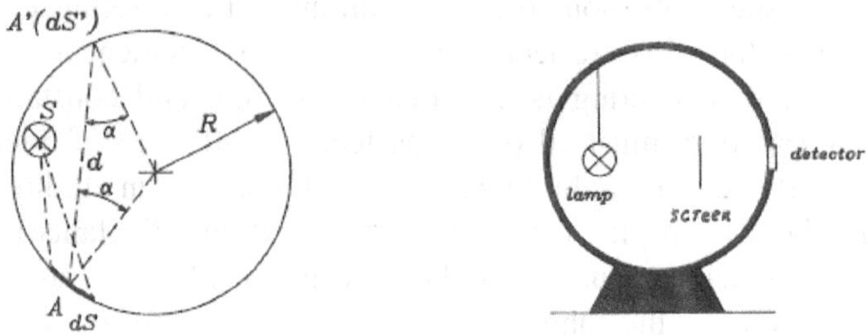

Figure 6.5. *Method of the integrating Ulbricht sphere*

In order for [6.21] to become [6.18], it remains to prove that E_r is constant in every point of the wall and thus that illuminance E_{rA} at A is independent of A. This illuminance is the integral of all the elementary luminous contributions coming from all points A' (of area dS') of the wall. Each contribution is expressed in term of luminance of this point according to its definition [6.2] "from the point of view of the observer" (even recalling that the elementary solid angle is the ratio between the transversal area placed at distance d of the origin of the cone confining this angle divided by the square of this distance). This justifies the third member of the following chain of relations:

$$E_{rA} = \int_S (dE_{rA})_{A'} = \int_S L_{A' \to A} (\cos^2 \alpha / d^2) dS' = \int_S L_{A'} \, dS' / 4R^2 \qquad [6.22]$$

In the passage to this last member, we have called on the property of isotropy of the wall to remove the arrow in the subscript of L because, for such a wall, $L_{A'}$ is independent of the direction considered. This ends the demonstration that E_{rA} is independent of A (= E_r). Expression [6.21] thus comes to [6.18], which shows that the reflection illuminance on any point of the wall is proportional to the searched flux, independently of this point and also of the source location. Furthermore, it should be noted that the size of the source does not intervene in any way in this reasoning. The measurement method therefore comes to place the head of a lux meter in an arbitrary point of the wall and to protect it from the direct sight of the source by interposition of a screen (Figure 6.5). The flux will be ideally deduced from the reading by [6.18].

Nevertheless, as was expected, the reality is not so attractive. The screen (also painted in white), the opening for the measuring cell, the lamp support and the lamp itself (eventually including its luminary) are all obstacles to the free play of reflections on the wall and the formula will have to be corrected by a factor k (*bulk factor*) lower than 1:

$$E_r = k\frac{\Phi}{4\pi R^2}\frac{\rho}{1-\rho} \quad [6.23]$$

The indirect illuminance is not more directly calculable by this formula because the bulk factor is practically undeterminable. Moreover, the reflectance is not strictly constant on the whole surface and also varies with time (tarnishing, dust).

The integrating sphere is therefore exclusively used in a *substitution mode.* The lamp under test (of flux Φ_T) and a reference lamp (of flux Φ_R) are successively introduced in the sphere and the measurements m_T and m_R at the wall lux meter (this does not require calibration, as its measure has only to be proportional to the illuminance) are respectively noted. The ratio of the fluxes is therefore proportional to the ratio of the readings:

$$\Phi_T = \Phi_R \frac{m_T}{m_R} \quad [6.24]$$

In practice, the measurement with the reference lamp is only made occasionally and serves for determining the *sphere factor* (SF):

$$SF = \Phi_R / m_R \quad [6.25]$$

The flux under test is now simply deduced from the formula

$$\Phi_T = SF * m_T \quad [6.26]$$

This formula supposes the identity of two parameters when the radiation is successively coming from the lamp under test and from the standard lamp:

– the same reflectance ρ of the painting. This is difficult to be respected when the spectra of both lamps are very dissimilar because the hypothesis of spectral non-selectivity of the painting cannot be strictly satisfied;

– the same bulk factor. A solution exists to free us from this constraint.

These two questions needing more consideration will be addressed in the following text.

A few more words on the used lux meter: the measuring cell must be cosinus corrected because it receives flux from the whole hemisphere facing the measuring head; furthermore, in the case of heterochromatic photometry (spectrum of the unknown different from the spectrum of the reference), this cell must of course be spectrally matched to $V(\lambda)$.

Comments on the wall reflectance

Expression [6.23] shows that in order to increase the measurement sensitivity, it is interesting to use a painting as white as possible (ρ approaching 1). If e.g. we pass from $\rho = 0.80$ to $\rho = 0.98$ (typical value for a painting based on magnesium oxide), the sensitivity (multiplying coefficient of Φ) will be multiplied by about 12. However, this is not without problems due to the incidence of the ρ variation (in time or as a function of λ) on the sphere factor as large as ρ is large. So, a reduction of 1% on ρ will reduce the sphere factor by 5% if $\rho = 0.80$ and by 50% if $\rho = 0.98$! We understand for the same reason that the spectral constancy of ρ is even more critical since the spectrum of the lamp under test deviates from this of the reference lamp used for the sphere calibration. The global reflectance ρ of any material is indeed dependent on the spectrum of the illuminant if the spectral reflectance $\rho(\lambda)$ is not constant as a function of λ.

The sphere theory lies on the hypothesis of isotropic diffusion of the wall painting. It is only under this condition that the play of successive reflections on the wall will make the (indirect) illuminance uniform on each point. Once again, this hypothesis is imperfectly respected with the consequence that this uniformity will be less realized as the source radiation is directional (as is often the case with LEDs). This effect will be more marked with a painting of low reflectance.

Measurement of bulky sources

Expression [6.24] assumes that the bulk factor k is the same in the presence of the reference lamp and in the presence of the unknown. This hypothesis is less respected the more both sources are dissimilar, namely in volume. To overcome this difficulty, the general method of the auxiliary lamp is appropriate [CIE 89]. This lamp is added in the sphere and is completed by additional screens masking it from the measuring window as well as from the main source, reference or unknown, in its presence. The sphere theory showed us that the successive reflections tend to make the (indirect) wall illuminance uniform. It is now pertinent to state that, for a given occupation of the space by the objects present in the sphere, the play of these successive reflections will be hindered independently of the initial origin of the light. In other words, the bulk factor will be practically the same when the sphere is lighted by the auxiliary lamp or by the main lamp if these two objects are simultaneously present during the measurements. This suggests the following method:

a) using the lamp under test as the main lamp and noting the measurements m_T and m_A by turning on the unknown lamp (T) and the auxiliary lamp (A) alternately. Expression [6.24] remains true by substituting subscript A with subscript R as the bulk factor k_T is the same for both measurements;

b) doing the same with the reference lamp (R) as the main lamp: measurements m_R and $m_{A'}$. Expression [6.24] is also correct by replacing subscript T with subscript A', because the new bulk factor k_R is the same for both measurements. Combining both situations, we finally obtain:

$$\Phi_T = \Phi_R \frac{m_T m_{A'}}{m_R m_A} \qquad [6.27]$$

The flux determination now requests the measurement of four illuminances. This relation indicates that the ratio $m_{A'}/m_A$, which is nothing other than the ratio of the bulk factors in both configurations, plays the role of a correcting factor for the result obtained by the application of the direct method. The auxiliary lamp method is only necessary if this ratio deviates from 1 which is actually a frequent occurrence. It should be noted that the knowledge of Φ_A which disappeared from [6.27] is not required, but only its stability during the full measurement process.

LED positioning in the sphere

Even if in principle the source position in the sphere is arbitrary, it is recommended, in order to reduce the disparity errors between the reference lamp and the unknown, to place the LEDs near the center of the sphere (Figure 6.6a). The screen will be at about halfway between the source and the detector. The eventual auxiliary LED will be near the wall facing the detector with its protecting screen nearby. In case of very directional LEDs (without emission at the rear), we can also place them near the wall, which facilitates their manipulation (Figure 6.6b).

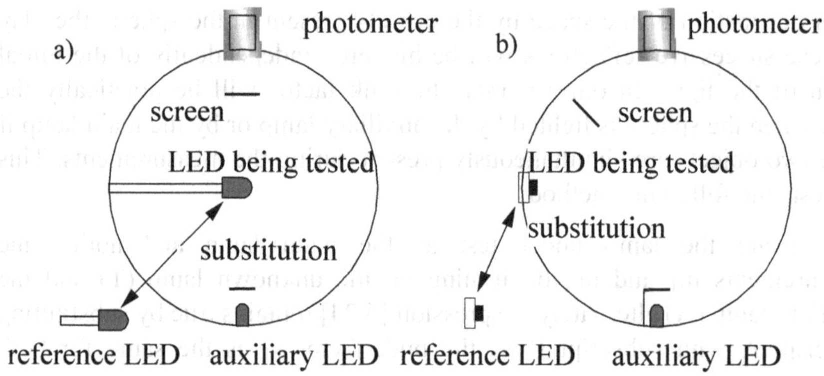

Figure 6.6. *Recommended geometries for the measurement of the total flux of LEDs in sphere: (a) for all types of LEDs, (b) for LEDs without rear emission*

The usual sphere diameters for LED measurements are between 20 and 50 cm. A large diameter favors the spatial uniformity of the wall illuminance (due to the lower relative bulk effect of the screen) and thus the measurement accuracy, but to the detriment of the sensitivity (which decreases according to the square of the diameter). It is worthwhile to note here that if sources of the same geometry (same bulk, same radiation diagram) are compared, it is not necessary that the cavity be spherical: a parallelepipedic box, for example, is sufficient.

Use of a spectroradiometer

A spectroradiometer can be substituted for a lux meter as a wall detector and the luminous flux of the LED can be deduced from the measurements if the spectroradiometer is in a mode measuring the spectral density of the energetic flux, as described in section 6.1.6.4. The difficulties mentioned above linked to the selective character of the painting and to the cell mismatch are in such a way overcome.

6.1.5.2. *Measurement of the "partial LED flux"*

The measurement of the partial flux of a lamp in a given solid angle assumes that the detector is placed at a distance sufficiently large to assume the source punctual. This is not always feasible for reasons of space. In the case of LEDs, and with the care of pragmatism and of uniformity of the experimental procedure already encountered for the intensity measurements, The CIE recommends the use of the notion of "partial LED flux". This requests that the measuring window be a circular opening of 50 mm diameter. The "partial LED flux" is defined as the flux propagating within a cone of specified angular opening, aligned along the LED mechanical axis and the apex of which is at the LED front extremity. For a plane angular opening of x degrees, the distance d from the cone apex to which the measuring window, centered on the cone axis, must be placed is given by the formula:

$$d = \frac{25}{tg\, x/2} \text{ [mm] } (0° \leq x \leq 180°) \qquad [6.28]$$

The symbol for this quantity is $\Phi_{LED,x}$, where x is the angular opening of the cone expressed in degrees. This quantity, which is rather arbitrary as explained, has the advantage of simplicity and reproducibility of measurements for any type of LED.

The measurement configuration of Figure 6.7, which conforms to this definition, is recommended. The screen is placed halfway between the opening and the measuring cell. It has to be as small as possible while completely masking the opening viewed by the cell. The space between the LED and the sphere opening must of course be protected from any parasitic surrounding light.

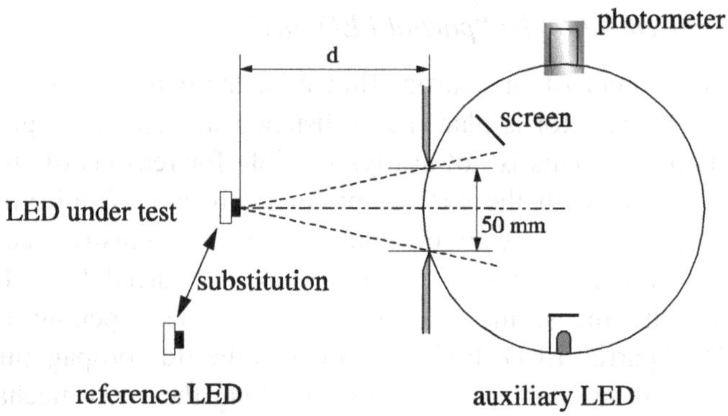

Figure 6.7. *CIE recommended geometry for the measurement of partial flux LED*

6.1.6. *Spectral measurements*

6.1.6.1. *Spectral distribution*

For any radiometric (or energetic) quantity X_e, its spectral distribution is the derivative of this quantity as a function of λ:

$$X_{e\lambda}(\lambda) = \frac{dX_e(\lambda)}{d\lambda} \qquad [6.29]$$

This quantity is determined with the help of a spectroradiometer. The core of this instrument is a monochromator, the output radiation of which would ideally have a negligible bandwidth for each wavelength. For LEDs, it is recommended (see below) not to exceed 5 mm (bandwidth at half power of the maximum). For the correct use of this instrument, see publication CIE 63 [CIE 84b]. It is always possible to come back to the global photometric (or luminous) quantity by application of formula [6.1]. This requires that the spectral distribution is known in absolute value and not only in relative value, as is often the case.

6.1.6.2. *Quantities linked to the spectral distribution*

The basic LEDs are mono-color: their spectrum presents one band only, which is bell-shaped. The main parameters are:

– the peak wavelength. When only the relative spectral distribution is sought, this is usually normalized towards this maximum, in such a way that its value is 1 at this point,

– FWHM (*full width at half maximum*) $\Delta\lambda_{0.5}$: bandwidth covering the band higher than half the maximum. This band is often narrow (from 30 to 50 nm).

The multiple LEDs in their different packaging (in particular, white LEDs) generally present a series of spectral bands shifted along λ. In the case of white LEDs using the principle of fluorescence, the spectrum has the shape of one or several narrow bands completed by a continuous spectrum.

6.1.6.3. *Colorimetric quantities determined from the spectral distribution*

The knowledge of the spectral distribution (its relative value is sufficient) allows the colorimetric parameters of the radiation to be known. It mainly concerns its two chromatic coordinates in any of its usual representations (e.g. x-y or u-v or u'-v') or, alternatively for mono-color LED, the dominant wavelength and the purity. For white LEDs, which have the property of being in the vicinity of the black body curve in the color triangle, the sole "correlated color temperature" parameter is sufficient. For these concepts and calculations, see publication CIE 15 [CIE 04].

The chromatic parameters of a source reveal nothing on the manner with which the color of the objects lighted by this source is rendered. This information is important to be known for white color sources. The pertinent parameter is here the *color rendering index* (*CRI*). The general measurement method evaluates the chromatic distortions of a reference chart of colored samples (of pastel hue) when they are successively lighted by the source under test and by a reference source, most often the illuminant A of the CIE (see publication CIE 13 [CIE 95]). Recent studies made on white LEDs called into question this index, which led the CIE to recommend starting a study aiming to propose a supplementary index which, without replacing the precedent one, should complete the information on this point (see CIE 177 [CIE 07b]). The question of colorimetric quantities is more widely considered in Chapter 7.

6.1.6.4. *Spectral measurements of LEDs*

The spectral distribution of an LED radiation can be measured with a spectroradiometer in four different modes: 1) irradiance mode, 2) total flux mode, 3) partial flux mode and 4) radiance mode. In irradiance mode, the radiation spectrum is determined in a particular direction while in a flux mode, it is averaged on a total or partial angular sector. The mono-color

LEDs (one narrow peak) generally present a good chromatic uniformity in every direction in such a way that the irradiance mode is here appropriate. At the contrary, the white LEDs often have an irregular color spatial distribution, therefore requiring a flux mode if their average color is sought. The radiance mode characterizes the spectral distribution of the radiance (i.e. the energetic luminance) of the emitting surface of the LED when this has a certain extension.

6.1.6.4.1. Irradiance mode

The LED under test is placed at a certain distance from the entrance opening of a spectroradiometer. It is important that this receives a uniform irradiance, which is obtained by interposition of a small integrating sphere or of a diffusing filter. This latter can be in transmission or reflection and combined with an optical fiber conducting the light from the output of the filter to the entrance of the spectroradiometer. Alternatively, this can be in radiance mode by directly aiming the filter using an appropriate optic. These setups will also be used for the instrument calibration with a reference lamp (usually a quartz-halogen tungsten lamp for which the spectral repartition is known) as for the measurement of the LED under test, each lamp being placed in the same position because it is only at this condition that the instrument is illuminated in an identical way in both cases, independently of the respective size of both sources.

6.1.6.4.2. Flux mode

In the spectral measurement of total flux, the geometry of the setup is identical to that of the sphere in measurement of total luminous flux (Figure 6.6), the difference being that the lux meter is replaced by a spectroradiometer. The whole setup is calibrated against a spectrally known reference flux lamp (the same tungsten lamp as above).

Spectral properties requested for the spectroradiometer

Each spectral measurement runs the risk of being inaccurate by reason of the non-zero bandwidth of the beam coming out of the spectroradiometer as well as of the sampling wavelength interval. Ideally, these quantities should be as small as possible, but the "cost" to pay is on the one hand a reduction in sensitivity and on the other hand an increase of the measuring time. This problem is particularly acute for a mono-color LED which presents a narrow emission band. The effect of the non-zero monochromator bandwidth is to enlarge the measured spectrum of the LED and therefore to alter the calculated chromatic parameters. As an example, a bandwidth of 10 nm (FWHM, triangular profile) runs the risk of provoking an error of 0.003 on the u'-v' chromatic parameters for a red or blue LED. These errors are proportional to the square of the bandwidth. They are thus reduced by a factor of 4 if a bandwidth of 5 mm is used, which is the highest recommended value. The sampling interval is less critical for chromatic parameter determination, but this quantity is important if the peak wavelength of a mono-color LED emission is sought. It is recommended to use a maximal sampling interval of 2.5 nm.

6.2. Electrical characteristics of LEDs

An LED is above all a diode and thus presents the electrical properties of this component. Operated with DC power applied in a forward bias direction, it has a forward voltage V_F depending mainly on the current going through it and on the junction temperature. The emitted light quantity is also a function of these two parameters. In order to obtain a correct measurement, it is thus essential to perfectly control the different electrical and thermal parameters of the diode.

6.2.1. *Forward voltage*

In an LED, the light is generated inside the material, at the junction of two semiconductors differently doped. The photons are emitted following the recombination of pairs of electrons-holes passing from one energetic level to another. This energy shift, and thus the energy of the generated photon, depends on the semiconductor materials used. By combining different materials, diodes emitting photons of different energy will be obtained, which corresponds to different wavelengths (inversely proportional to this energy by Planck's law), in other words to lights of different colors.

Supplied in forward direct current, the LED presents, for a given junction temperature and current, a voltage depending on this energy shift (Figure 6.8: λ decreases from (a) to (e); e_g = "gap energy"). We will thus have voltages V_F different as a function of the color emitted by the LED. The relation between the current I_{LED} crossing the diode and the voltage V_F is exponential:

$$I_{LED} = I_{saturation}\left(e^{\frac{eV_F}{kT}} - 1\right) \quad [6.30]$$

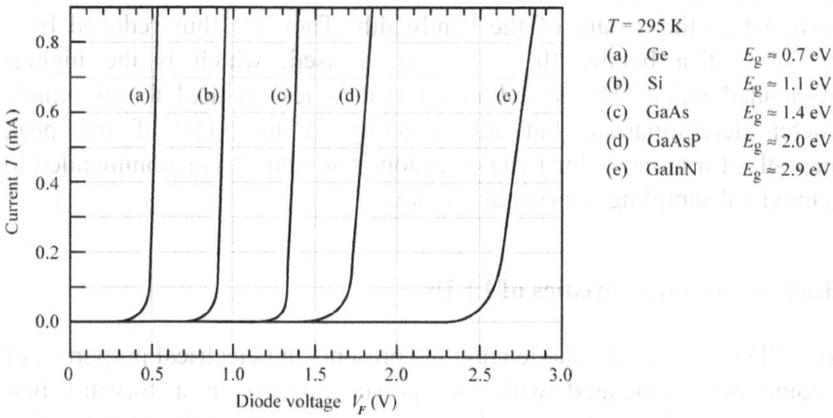

Figure 6.8. *Relationship between current and voltage for different types of diodes [SCH 06]*

Considering the curves of Figure 6.8, we observe that a slight voltage variation around its steady-state value will induce a large current variation and thus a large variation of the light quantity emitted by the LED. It is therefore very difficult to control the LED by way of a voltage source. A current source will thus be used to supply the LED and this current will be carefully measured since it defines the steady-state point and is responsible of the light flux emitted which depends quasi-linearly on this point in normal operating conditions.

6.2.2. *Temperature effect*

Let us correct the erroneous messages such as "the consumption of an LED is practically zero" or "an LED does not produce heat" which are circulating amongst the general public. Concerning the consumption, we know that the most pertinent parameter for a light source is its luminous efficacy (in lm/W). Even if it is true that the LED performances are here better than those of an incandescent lamp, the values of fluorescent lamps (TL tubes) are not yet achieved, though it is the long term objective. Concerning the thermal aspect, the LEDs (for the visible applications) are negligible and it is on this point that we can declare that they do not produce heat (in the sense that their light does not produce any heat). It is, by the way, for this reason that LEDs are more and more used in art galleries (also for the additional motive that UVs are absent in this radiation). On the contrary, the element itself can reach important temperatures in such a way that, in practice, a correct thermal management is imperative. Let us number this apparent paradox due to the absence of infrareds by comparison with traditional incandescent lamps. In the case of a traditional incandescent lamp, only 5% of the consumed electrical energy is transformed into visible light and 75% of the radiation is emitted in the infrared. The remaining 20% are transformed into heat and dissipated by conduction via the cap, or by convection in the air. For the LED, more electrical energy is transformed into visible light (about 20%), but without other emissions. The remaining 80% are therefore transformed into heat inside the material on a punctual manner and have to be extracted by conduction outside the LED at the risk of a dangerous temperature increase inside the element.

The junction temperature influences emitted light quantity (see Figure 6.9), but also the voltage V_F and thus the color.

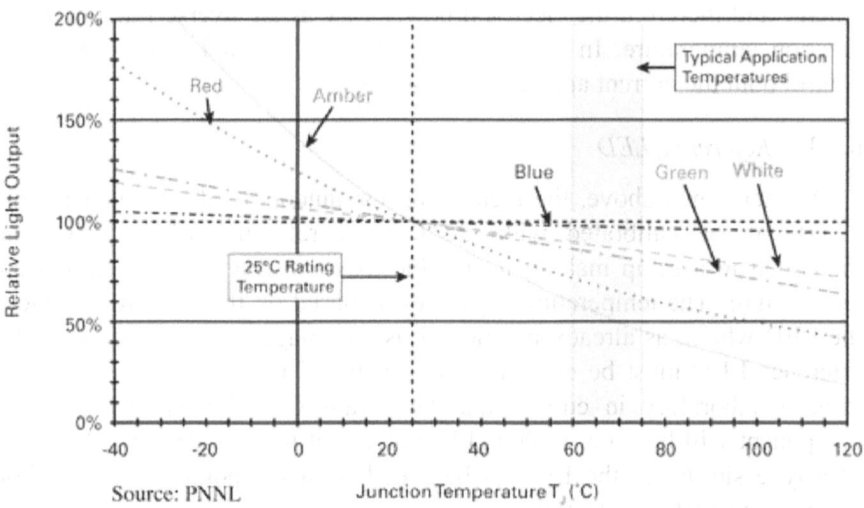

Figure 6.9. *Temperature effect on the luminous flux [DOE 07]*

The temperature coefficient of V_F is negative and amounts to about

$$dV_F / dT_j = -1.5 \text{ to } -2.5 \text{ mV/K} \quad [6.31]$$

The peak wavelength emitted by the LED will also slightly vary as a function of the junction temperature:

$$d\lambda_p / dT_j = 0.1 \text{ to } 0.3 \text{ nm/K} \qquad [6.32]$$

This will have two effects: the first, quite naturally, on the color of the light and the second on the light efficacy of the LED. Indeed, for blue and red LEDs, a small shift of the peak wavelength will significantly modify the flux value because this wavelength is at a location where the slope of $V(\lambda)$ is important (see Figure 6.1). Let us still mention that an increase of the junction temperature reduces the live duration of the LED.

6.2.3. *Operating conditions of LEDs for photometric measurements*

For numerous traditional light sources, a strong correlation exists between the luminous flux and the consumed electrical power, which makes it possible to control the flux regime by way of this power. This is less pertinent for LEDs because for constant current and thus constant flux, the voltage and therefore the electrical power vary as an inverse function of the junction temperature. In order to stabilize the flux, it is thus essential to control both the current and the temperature.

6.2.3.1. *Reference LED*

As mentioned above, the measuring instruments used for characterizing LEDs can be calibrated with the help of reference LEDs. These are integrated in a setup making it possible to work at a constant current and temperature. The temperature regulation is based on the forward voltage of the LED which, as already mentioned, is the image of its temperature. The reference LED must be calibrated by a national metrological institute or a primary laboratory in current and temperature conditions for which the component will be used. It should be noted that the reference LED must be of a type similar to the LED to be tested from the point of view of both spectral and spatial distribution.

6.2.3.2. *Steady state DC operation*

Without other specifications, the LED is measured at an ambient temperature of 25°C. The energy dissipated in the junction will increase the temperature of the component until a stabilization point is reached. This thermal equilibrium point results from the heat transfer between the LED and the ambient air. This state will depend on a large number of parameters, such as the manner in which the component is fixed on its support, the size of this, the size of the supply wires, etc.

6.2.3.3. *Pulsed operation*

In order to be free of the thermal stabilization problem of the LED, another technique exists, consisting of feeding the LED with short periodic pulses at the nominal current. If the duty cycle (the time ratio between the "on" state and the period of the pulse) is short (typically 0.1%), we can consider that the heat dissipated during the "on" phase will not influence the junction temperature which is thus the ambient temperature in which the LED is immersed. By applying this supply mode to an LED placed inside an oven brought at different temperatures, it is possible to determine the relation between the junction temperature and the forward voltage V_F already pointed out above. Afterwards, at the time of photometric measurements, it will be sufficient to invert this relation in order to know, from the measured value V_F, the junction temperature in effect during this photometric test.

It is important to mention here that the flux of an LED supplied by short impulses at an ambient temperature of 25°C does not correspond to the flux that the LED will produce when placed inside a luminary. Indeed, the typical operating temperature of an LED in actual applications is often higher than 60°C, which induces a decrease in luminous efficacy (Figure 6.9).

Let us also note that the voltage V_F must be measured by wires separated from the supply wires and directly connected to the LED contacts (four-pole connection). In such a way, the voltage drop in the supply wires is bypassed and, consequently, the measurement of V_F which must be known with high accuracy, as already mentioned, is not disturbed.

6.2.4. *Stand of the normalization*

The actual recommendations mentioned in [CIE 07a] do not impose precise operating conditions from electrical and thermal points of view. The DC and pulsed supply methods are not described in detail, which allows a great freedom of interpretation. This does not favor the performances comparison between manufacturers.

Inside the CIE, a technical committee was recently formed with the aim of determining a practical method for controlling current and temperature making it possible to measure the performances of the LEDs in conditions near their normal functioning.

6.3. Bibliography

[CIE 82] "Methods for characterizing the performance of radiometers and photometers", CIE 053-1982.

[CIE 84a] "Determination of the spectral responsivity of optical radiation detectors", CIE 064-1984.

[CIE 84b] "The spectroradiometric measurement of light sources", CIE 063-1984.

[CIE 87a] "International lighting vocabulary – Vocabulaire international de l'éclairage", IEC/CIE 017.4-1987.

[CIE 87b] "Methods of characterising illuminance meters and luminance meters", CIE 069-1987.

[CIE 87c] "The measurement of absolute luminous intensity distributions", CIE 070-1987.

[CIE 89] "Measurement of luminous flux", CIE 084-1989.

[CIE 95] "Method of measuring and specifying colour rendering properties of light sources", CIE 013.3-1995.

[CIE 04] "Colorimetry, third edition", CIE 015-2004.

[CIE 05] "Photometry – the CIE system of physical photometry", CIE S010/E:2004/ISO 23539 : 2005(E).

[CIE 07a] "Measurement of LEDs", CIE 127-2007.

[CIE 07b] "Colour rendering of white LED light sources", CIE 177-2007 .

[DOE 07] *US Department of Energy*, "Thermal management of white LEDs", 2007 (on the Web).

[SCH 06] Schubert E.F., *Light-Emitting Diodes*, second edition, Cambridge University Press, 2006. http://www.lightemittingdiodes.org.

Chapter 7

Quality of White Light from LEDs

7.1. Introduction: white light and visual quality

What are we expecting from artificial light? To provide visual performance and comfort for the end user while emphasizing his life setting and environment in acceptable economic conditions. Technically it means: high efficiency, low power consumption, controllable intensity, possibly lack of ultraviolet and infrared, and more generally, "white light" in "sufficient quantity" and "visual quality".

To fulfill these requirements, the light manufacturer has quantitative data on electrical consumption, maintenance needs, sources emission and spectrum of the light emitted. Here we will focus on the spectrum of the light emitted by LEDs, and for the interest of the user who desires a good quality of light for his visual environment.

7.1.1. *White light*

Typically, white light mimics the natural daylight under which we constantly live and which delivers a full spectrum that we are used to. Daylight does change depending on the hour of the day, the cloud cover, sunlight and the season, but its margins of variation are strictly limited and

---

Chapter written by Françoise VIÉNOT.

its spectrum practically does not vary from the average spectrum profile (Figure 7.1).

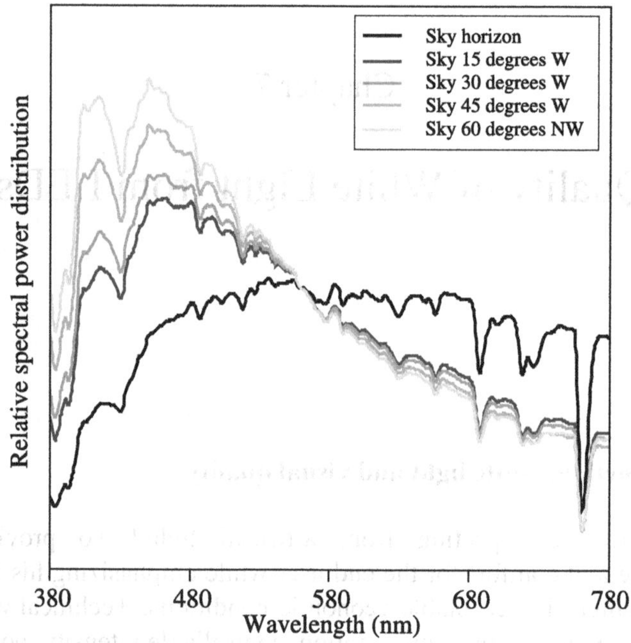

Figure 7.1. *Relative spectral power distribution of daylight. Measurements normalized at 560 nm. Sky measurements above the horizon, facing west, 20 December 2007 above Paris*

The result of this is that natural daylight can seem gold or pink tinted, and in this case considered as "warm", or neutral to slightly blue, and felt as "cool". In order to accurately characterize the color of white light, light makers have introduced a scale known as "color temperature". This scale solely reflects the color of the light, but not its spectrum or quality.

7.1.2. *A few ideas on the quality of light*

From our experience, not all artificial light sources provide the same visual quality. Quality can be assessed following different perspectives. It can be to guarantee color rendering, as in the dye industry; in that case, the Color Rendering Index (CRI) does provide enough information. It can also be to faithfully discriminate the true colors of a masterpiece painting in a museum, to render a natural look for the cosmetic industry, to bring eye comfort at dusk, to ensure visual performance on a detailed task or for general well being.

Visually, it is the observer's judgment which will assess the quality. For that very reason, the light maker needs an appropriate toolbox to predict that judgment.

7.1.3. *The human visual function: receptors, retina, brain*

Color has its origin in the variations of the spectrum of the light reflected by the different materials. These variations are detected by the human eye and then perceived as colors. The eye does not accurately analyze the spectrum; it converts light into signals that the brain then interprets. In a more detailed way, light entering the eye forms an image on the retina. It is first detected by the retinal cells: three families of cone cells sensitive to different spectra in the visible range. Only wavelengths ranging from 360 to 830 nm can be detected by the cones.

The signals generated by the cones are then compared in the retinal neurons before being sent to the brain (Figure 7.2). The management of color differences and contrasts of the image is done by this neural networks. It has generally been agreed since the 1980s that visual information travels in the brain through two main paths: one for object recognition and the other for localization and motion. In the first path, the brain compares the information from the observed object with references also present in the scene or memorized and finally attributes a color to the object. This path is used for conscience representation of our environment, a second path is used for motion reflex control.

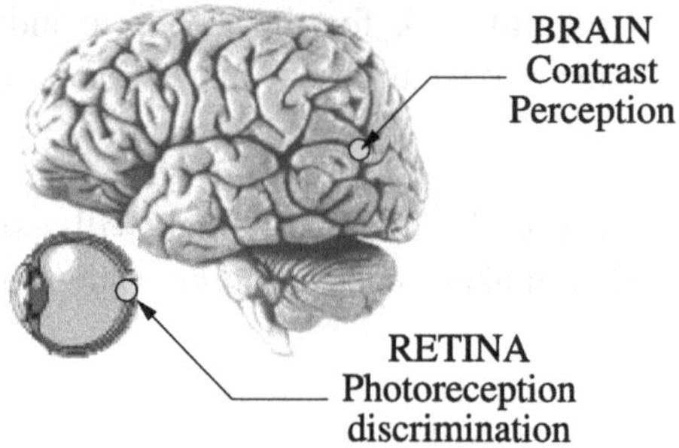

Figure 7.2. *Signal treatment in the visual system*

In the sensorial domain, color is expressed in a three-dimensional space in terms of chromaticity (red, yellow, green, blue), brightness (light, dark) and saturation (Figure 7.3).

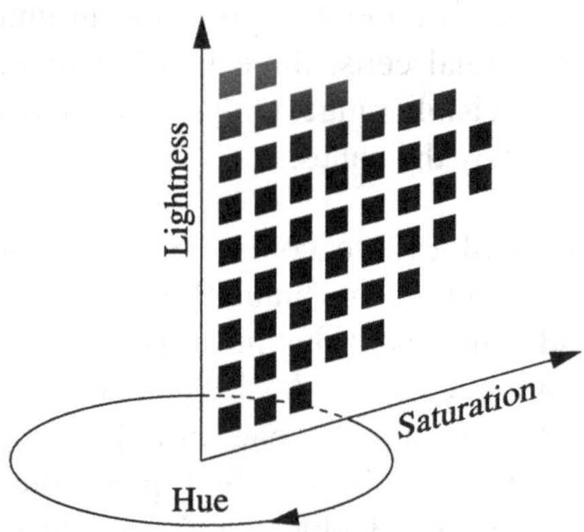

Figure 7.3. *Three dimension representation of the colors following the Munsel classification (see color plate section)*

7.1.4. *Chapter presentation*

We will first present the colorimetric methods that make it possible to accurately specify the color of the light and to assign a CRI. Then we will discuss the questions arising from the definition of color rendering applied to LEDs.

7.2. Notions of colorimetry and photometry

7.2.1. *Colorimetry*

Colorimetry refers to a gathering of data and methods dedicated to color specification, differentiation and appearance [CIE 04].

A color is specified by tristimulus values X, Y, Z. This specification, linked to the light absorption by the three families of cones in the retina, has its origins in visual experiments. Experience shows that it is possible to reproduce all color stimuli (luminous ray reaching the human eye) by mixing the three stimuli known as primary colors. Experience also shows that color matching has additivity properties. Finally, we usually interpret the luminous characteristic – dark, light – of the light separately from its chromatic feature – red, yellow, etc.

We can therefore apply to color the laws of linear algebra and represent a color by a vector in three-dimensional space. There are many different color spaces, such as RGB for television or the XYZ chromatic space recommended by the Commission Internationale de l'Eclairage (CIE) (Figure 7.4). The chromaticity of a color is given by the relative values of the tristimulus values X, Y, Z and is represented in two dimensions in the chromaticity diagram x, y (Figure 7.5).

$$x = \frac{X}{X+Y+Z} \; ; \qquad y = \frac{Y}{X+Y+Z} \qquad [7.1]$$

The chromaticity diagram x, y does not carry any information on the luminance of the color. It is given directly by the trichromatic value Y.

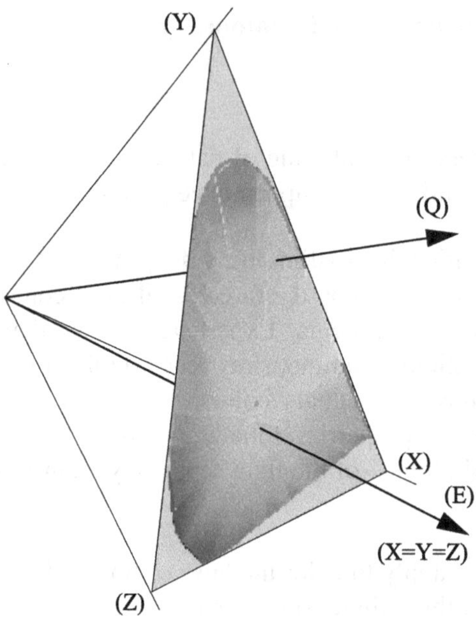

Figure 7.4. *Representation of colors in the color space XYZ. The equi-energy white point is at the center of the diagram (see color plate section)*

To evaluate the visible differences between two colors, uniform color spaces are used, where at all points almost equivalent color differences are represented by identical distances.

The CIELAB uniform color space is generally preferred for specifying the material color. The coordinates L^*, a^*, b^* of the colors are obtained by a non-linear transformation of the tristimulus values X, Y, Z. In particular, the psychometric lightness L^* is a non-linear transformation of the luminance Y. The white material is the origin of the a^* b^* diagram.

The CIELUV uniform color space is useful to specify the colors of light sources. The light makers often study the colors of the sources independently of the emitted flux. The chromaticity is measured in the chromaticity diagram u', v', which is a projection of the chromaticity diagram x, y (Figure 7.6).

Recently, the CIE has introduced the CIECAM color appearance models to predict the appearance of a color replaced in its context.

7.2.1.1. *The color temperature*

The light is white with reference to natural daylight. Because the natural light follows a variation law that is strictly defined, its color and spectrum are closely related. The color variations of the daylight are almost identical to those of the "blackbody", also called the "Planckian radiator".

The blackbody is a theoretical radiation source that has the property to emit back all the received energy. The physical laws which predict the radiation are well known. Because the radiation only depends on the temperature, knowing the temperature makes it possible to know its emission spectrum and therefore its color.

By comparing the color of the natural light with that which would theoretically be emitted by the blackbody at different temperatures, it is possible to build a scale of colors for the white light sources. All white light sources have a *correlated color temperature* (*CCT*) expressed in Kelvin (scale of absolute temperature, which varies from the Celsius temperature scale such that 0°C = 273 K).

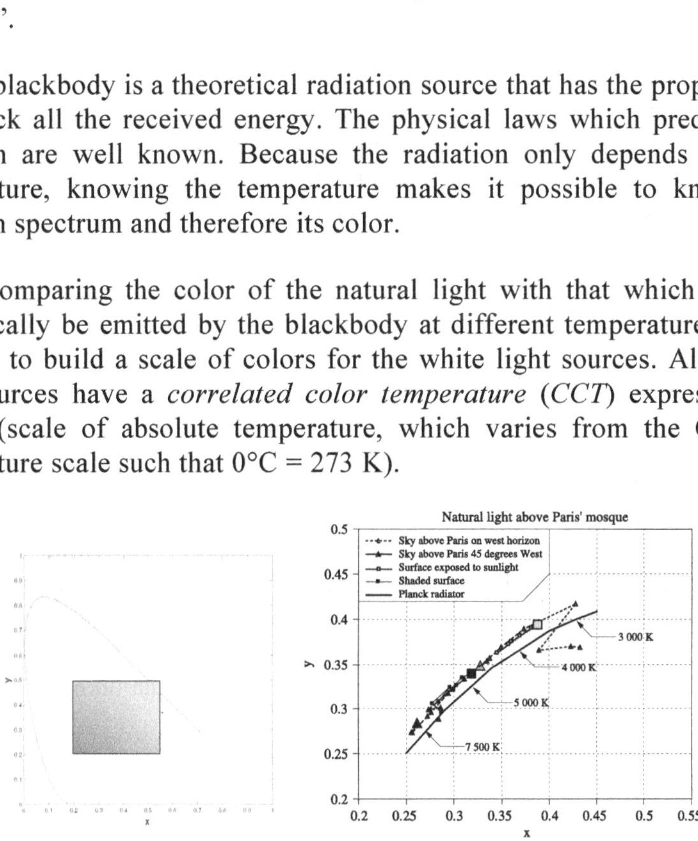

Figure 7.5. *Localization of the natural daylight on the chromaticity diagram x, y of the CIE (see color plate section)*

Light source	Correlated color temperature (CCT)
Warm	≤ 3,300 K
Neutral	3,300 K ≤ T_{pc} ≤ 5,300 K
Cool	> 5,300 K

Table 7.1. *Correlation of the color temperature with the light source [AFE 06]*

For example, a standard incandescent lamp emits white light with a CCT close to 2,500 K; direct sunlight has a CCT close to 5,000 K; the overcast north sky has a CCT close to 6,500 K; "warm white" fluorescent tubes emit light with a CCT of 3,000 K and "cool daylight white" fluorescent tubes emit light with a CCT close to 4,000 K. It should be noted that warm colors correspond to low correlated color temperatures, whereas cool colors have high CCTs.

7.2.1.2. *Natural daylight spectral model*

In colorimetry [SCH 07], the phases of daylight have been well described by Judd and his co-workers in 1964:

1) Chromaticity of the daylight phases follows a parabolic curve very similar to the Planck radiation spectrum in the chromaticity *x, y* (Figure 7.6). It is linked to the CCT.

If the CCT is between 4,000 K and 7,000 K, the equation is:

$$x_D = \frac{-4.6070\times10^9}{(T_{cp})^3} + \frac{2.9678\times10^6}{(T_{cp})^2} + \frac{0.09911\times10^3}{(T_{cp})} + 0.244063 \qquad [7.2]$$

If the CCT is above 7,000 K and below 25,000 K, the equation is:

$$x_D = \frac{-2.0064\times10^9}{(T_{cp})^3} + \frac{1.9018\times10^6}{(T_{cp})^2} + \frac{0.24748\times10^3}{(T_{cp})} + 0.237040 \qquad [7.3]$$

Thus, we can deduce the value:

$$y_D = -3,000\, x_D^2 + 2,870\, x_D - 0.275 \qquad [7.4]$$

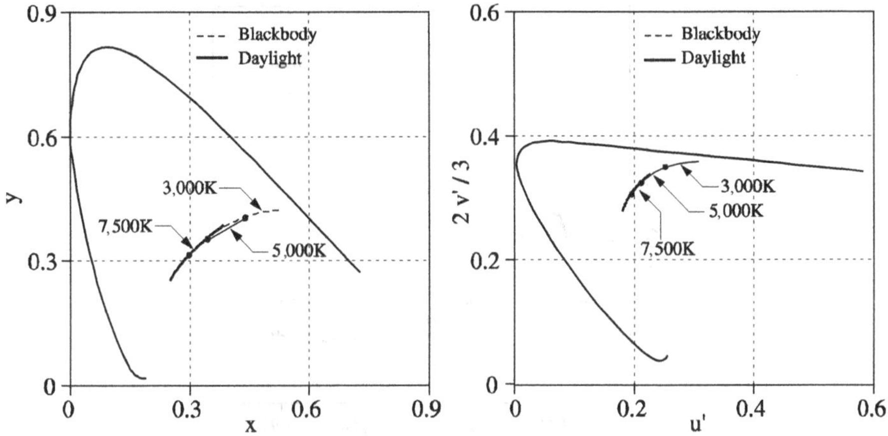

Figure 7.6. *Planckian locus and daylight locus in the x, y chromaticity diagram (left) and in the u, v uniform chromaticity diagram (right)*

2) The spectral distribution of the energy phases of the daylight can be described by three main components $S_0(\lambda)$, $S_1(\lambda)$ and $S_2(\lambda)$ following (Figure 7.7):

$$S(\lambda) = S_0(\lambda) + M_1 S_1(\lambda) + M_2 S_2(\lambda) \qquad [7.5]$$

where M_1 and M_2 depend on the trichromatic coordinates of the daylight phases and are given by:

$$M_1 = \frac{-1.3515 - 1.7703 x_D + 5.9114 y_D}{0.0241 + 0.2562 x_D - 0.7341 y_D} \qquad [7.6]$$

$$M_2 = \frac{0.0300 - 31.4424 x_D + 30.0717 y_D}{0.0241 + 0.2562 x_D - 0.7341 y_D} \qquad [7.7]$$

Tables give values for the vectors $S_0(\lambda)$, $S_1(\lambda)$ and $S_2(\lambda)$ each 10 nm.

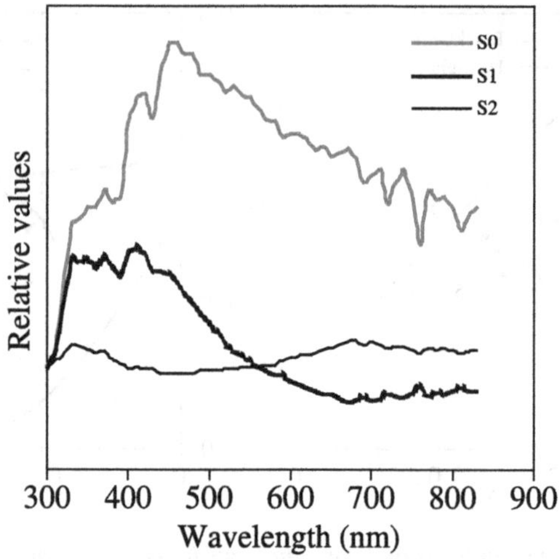

Figure 7.7. *Base vectors $S_0(\lambda)$, $S_1(\lambda)$ and $S_2(\lambda)$ for the daylight phase calculation*

7.2.2. *Photometric quantities*

7.2.2.1. *Definitions*

The human eye is sensitive to monochromatic radiation ranging from 360 to 830 nm, its maximum sensitivity being around 555 nm. The CIE has normalized the spectral sensitivity function or luminous efficiency[1] function for daylight radiation $V(\lambda)$, which makes it possible to convert the energy flux $\Phi_{e,\lambda}(\lambda)$ of a radiation, expressed in $W.nm^{-1}$, into luminous flux Φ_v or Φ (the v index is not compulsory) expressed in lumens (Figure 7.8).

$$\Phi = 683 \int_{360nm}^{830nm} \Phi_{e,\lambda}(\lambda) V(\lambda) d\lambda \qquad [7.8]$$

1. Here the word "efficiency" means the ratio between the luminous flux to the radiated energy flux or radiated power, not to be mixed up with "efficacy", which sometimes expresses the ratio of the luminous flux emitted by a source to the consumed electrical power.

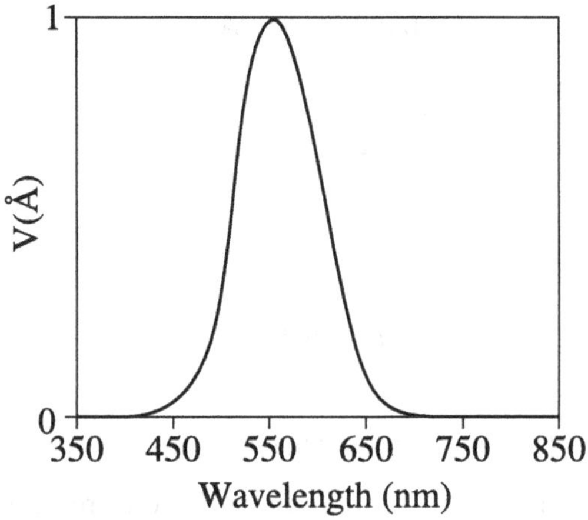

Figure 7.8. *Spectral luminous efficiency function V(λ) of the CIE standard photometric observer*

From the luminous flux, we can define the luminous intensity, the luminance and the illuminance (Figure 7.9).

The luminous intensity I_v or I in a given direction measures the luminous flux emitted from a point source into the solid angle of 1 Steradian. The unit is the candela (cd):

– the luminance L_v or L of a surface source in a given direction measures the ratio between the luminous flux emitted into the solid angle of 1 Steradian to the projected surface area in that direction. It is measured in candela per square meter ($cd.m^{-2}$);

– the illuminance E_v or E for a given surface is the luminous flux incident per unit area. The unit is the lux (lux = lm/m^2).

The equations defining the luminous quantities are given in section 6.1.1.

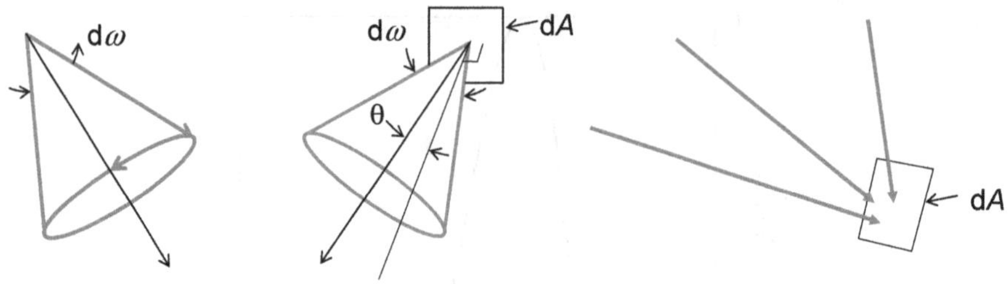

Figure 7.9. *Definition of intensity (left), luminance (center) and illuminance (right)*

7.2.2.2. *Figures of merit and recommended quantities*

For human vision, the useful light is the one reaching the retina. The illuminance at the retina's surface is precisely proportional to the luminance of the objects or the sources in the environment.

The visual functioning regime varies between night (scotopic vision) and day (photopic vision) (Figure 7.10). Color vision is only correct at daylight.

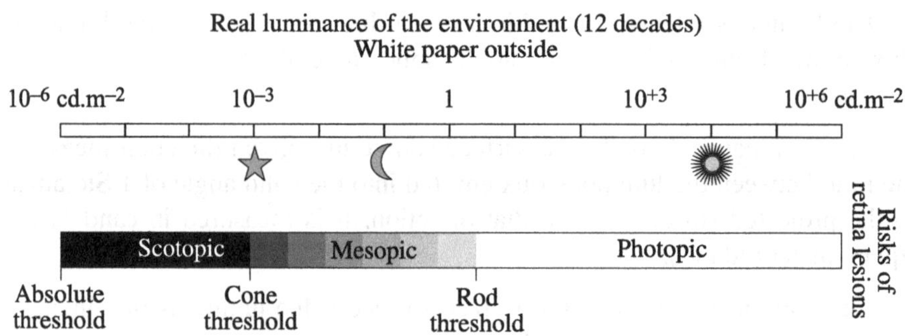

Figure 7.10. *Ranges of vision regimes*

Tables 7.2, 7.3 and 7.4 give some of the usual values of recommended illuminance and luminance.

Object receiving the light	Recommended illuminance (lux)
Fragile museum objects	50
School desks	500
Fine handworks	1,000
HDTV shooting, horizontal lighting on the ground	2,000

Table 7.2. *Recommended illumination values*

Diffusing object	Luminance ($cd.m^{-2}$)
Minimum perceptible	10^{-6}
White surface at full moonlight	10^{-1}
Curb with streetlights by night	1
Lower limit for reading	10
Video display with light background White surface with in-house lighting	1 to 5×10^2
Concrete at overcast weather	10^3
Sand with sunlight	10^4
Snow with sunlight	10^5

Table 7.3. *Typical values of luminance encountered in everyday life*

Sky	Luminance ($cd.m^{-2}$)	Chromacity coordinates x	y	Color temperature (K)
Light blue	600 to 4,000	0.247 0.262	0.251 0.270	30,000 15,000
Partially cloudy	1,000 to 4,000	0.279 0.294	0.291 0.309	10,000 8,000
Overcast	2,000 to 5,000	0.313	0.329	6,500

Table 7.4. *Typical luminance and color temperatures of different sky situations [WYS 82, extract from Table 1.2.8]*

7.2.2.3. *Risks of glare*

The luminance of a very small and directional point source can reach very high values, even if it emits a weak luminous flux. The risk of discomfort glare does exist with LEDs. The reflectors of lighting systems must prevent this.

Glare first induces a feeling of discomfort produced by shining surfaces – lamps or reflections – located in the visual field, without necessarily affecting the observer's visual capacities. In order to control the discomfort glare, it is recommended to follow the UGR (*Unified Glare Rating*) evaluation method [CIE 02a; AFE 06].

The disturbing glare, or disability glare, occurs when a luminous veil superimposes itself on the scene and significantly reduces the contrast of objects located near the point source. The luminance L of the veil due to a source with illuminance E in the eye plane depends of the angle difference θ between the source and the direction of sight. The disturbance worsens with the age A of the observer. The CIE proposes a formula valid for angle differences θ between 0.1° and 30° [CIE 02b]:

$$\left[\frac{L}{E}\right] = \frac{10}{\theta^3} + \left[\frac{5}{\theta^2}\right] \cdot \left[1 + \left(\frac{A}{62.5}\right)^4\right] \quad [7.9]$$

The exposure to blue light also presents a risk of photochemical degradation of the retina. The action spectrum $B(\lambda)$ of this risk reaches its peak at 440 nm (Figure 7.11). For a source with spectral radiance $L_e(\lambda)$ seen at 25 cm (for the definition of spectral radiance, see Chapter 6), the allowed observation time t is given by the equation of the *American Conference of Governmental Industrial Hygienists* (ACGIH):

$$t\sum_{400}^{740} L_e(\lambda)B(\lambda)\Delta\lambda \quad \leq 100 \text{ Joules.cm}^{-2}.\text{sr}^{-1} \qquad [7.10]$$

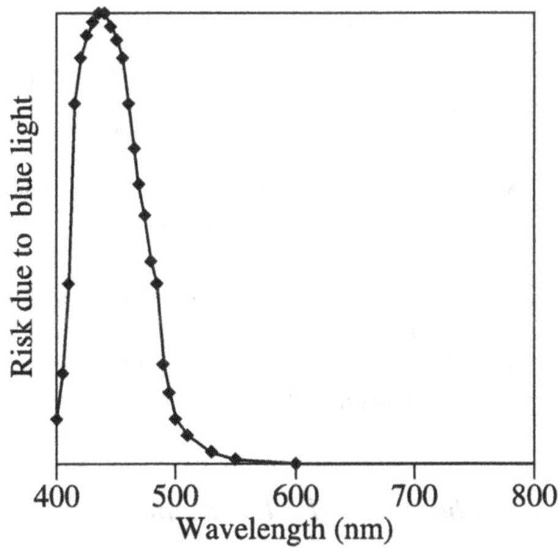

Figure 7.11. *Spectral function for photo-damage risk of the retina (from ACGIH 1985, 1989)*

7.3. Obtaining white light with LEDs

The three methods used to emit white light with LEDs are presented in section 1.3.1 in Chapter 1. Here we are interested in the spectrum of the emitted light which depends on the chosen method and conditions the color rendering.

7.3.1. *White light diodes based on short wavelength emission*

The white light is obtained by mixing the blue light emitted by a short wavelength diode and the yellow light emitted by fluorescence of a phosphor. The spectrum of the light emitted exhibits a narrow blue peak and a larger yellow peak. By controlling the deposition of fluorescent powder, the diode maker can adjust the color temperature. The spectrum is well covered. In 2007, a warm white LED was able to emit flux of 150 lm with a good CRI ($R_a = 80$) (Figure 7.12).

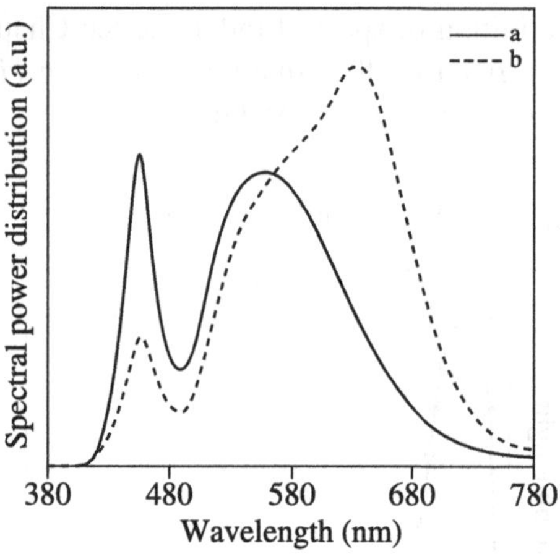

Figure 7.12. *Spectral energy distribution of white LEDs based on blue LEDs: a) cool white LEDs with one phosphor, correlated color temperature* T_{cp} *of approximately 5,000 K, CRI* $R_a > 70$*; b) warm white LEDs with several phosphors, correlated color temperature* T_{cp} *around 2,750 K ,CRI* $R_a > 80$

7.3.2. *White light diodes based on the UV diode*

The white light is obtained by mixing the colors emitted by the fluorescent powders in three distinctive regions of the spectrum. By choosing and controlling the fluorescent powder deposition, it is possible to adjust the color temperature. The CRI is improved for all color temperatures.

7.3.3. *Combining red, green and blue*

The combination of red, green and blue lights, in appropriate amounts, makes it possible to cover almost the full colored lights palette. By adjusting the intensity of three red, green and blue diodes, it is therefore possible to obtain a large range of color temperatures. The colorimetric calculations make it possible to estimate the assembled intensities of the LEDs to reach the color of light required by the user. The choice of red, of green and of blue is very wide. Although slightly higher than the efficiency of a white LED, the low luminous efficiency of the green diode limits the efficacy of this method. The spectrum of the light obtained displays three peaks of different intensities, and also important gaps within (Figure 7.13). The CRI is poor ($R_a < 60$).

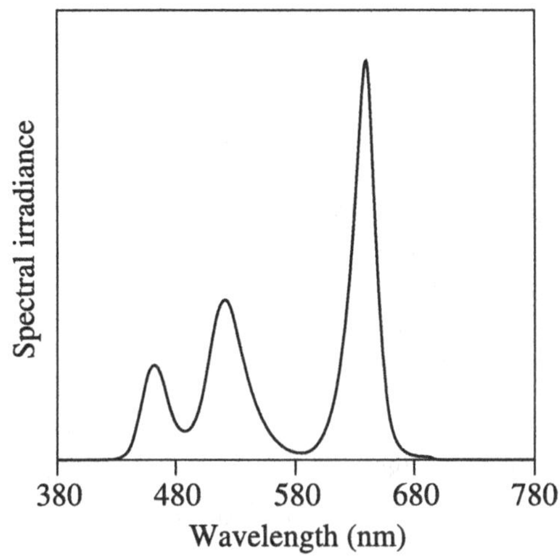

Figure 7.13. *Luminance spectral distribution of RGB combination of red, green and blue LEDs, correlated color temperature T_{cp} around 5,000 K, poor CRI $R_a = 22$*

7.3.4. *Examples of combining many LEDs, spectrum optimization*

It is possible to adjust the chromaticity of the light on the daylight locus and to optimize the final color rendering by assembling a large number of different LEDs (Figure 7.14). This is the case for the combination "red, amber, green and blue", or for the "red, green, cool white blue", which improve the color rendering compared to the simple "red, green and blue" combination [SPE 06]. When associating several types of diodes, white and colored, the color rendering can be excellent (Figure 7.14) [VIE 07]. With a white InGaN diode and yellow phosphor, and seven colored diodes, it is possible to accurately reproduce the spectral power distribution of natural daylight, which prevents the bad effects of metamerism [LI 06]. The temperature and time stabilities can, however, be problematic and require color readjustments (see sections 1.3 and 6.2.2 on temperature effects).

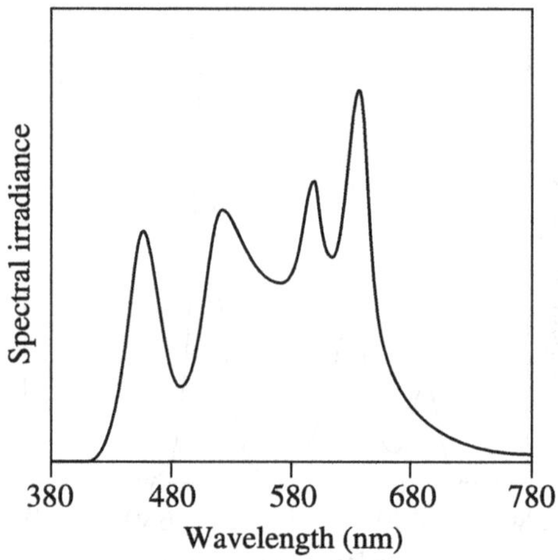

Figure 7.14. *Spectral distribution of a WWRGBA combination of one phosphor warm white, cool white, red, green, blue and amber diodes, optimized to give a color temperature of 4,000 K and a high CRI $R_a = 97$*

7.3.5. *Normalization of the color of white diodes*

LEDs are classified by their manufacturers with relation to tolerance "binning boxes", consisting of small color quadrangles which cover the chromaticity diagram at the near vicinity of the daylight locus, for a wide range of color temperatures (Figure 7.15). It is known as "binning" (see Chapter 5). The limits of the quadrangles are parallel to the Planckian locus and at the lines of equal color temperature. The surfaces of these boxes are uniform in the *u'v'* chromaticity diagram. Inside a given box, the colors of the diodes only vary by a few units of CIELUV color differences. Such variations are generally acceptable as long as the ranges of color differences do not spread for more than a few degrees of the visual angle.

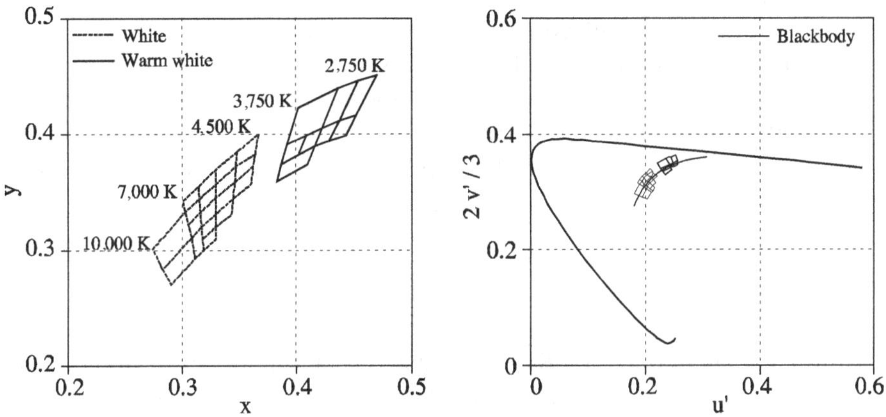

Figure 7.15. *Examples of color tolerance quadrangle organization for LED sorting. Left: limits in the x,y chromaticity diagram x,y. Right: limits in the u,v chromaticity diagram and the Planckian locus (from Lumileds, 11/07)*

For normalization and recommendations, we should refer to the following organizations:

– The CIE (www.cie.co.at) published the reference technical report on the CRI [CIE 95]. The CIE also published a recommendation on the methods of measuring the flux emitted by LEDs, and in particular measuring the spectral flux [CIE 07a], and finally the technical report on the need for a new light quality index for LEDs.

– The Illuminating Engineering Society of North America (IESNA) (www.iesna.org) published a technical report on sources and systems based on diodes [IES 02]

– The American National Standard Institute (ANSI) (www.ansi.org) published a recommendation on chromaticity tolerances of white LEDs depending on the color temperature [ANS 07].

7.4. Color rendering of sources

The CIE defines the color rendering as the effect of a light source on the color appearance of the objects exposed to its light, as compared to their aspect under a reference source [CIE 87].

The idea behind the characterization of the color rendering is unique. Starting with a given collection of colored samples, chosen either for their universality or conversely for their specificity, the aspect of these samples under the studied source is compared to the one they would have under a theoretical source or a reference source. This concept, applied for the definition of the CRI, can also be applied to all definitions of the visual quality: color discrimination and color appearance.

7.4.1. *The CRI of the CIE*

The CIE has defined a CRI which enables each light source to evaluate its quality with reference to natural daylight. Today, the generally used index is given by a recommendation of 1974, republished in the CIE 13.3-1995 [CIE 95]. The suggested method relies on the following principle (Figure 7.16). For a given sample, the color under the tested source is compared to the one it would have with a reference light. The CRI is given by the calculation of the color differences. It decreases when the differences increase.

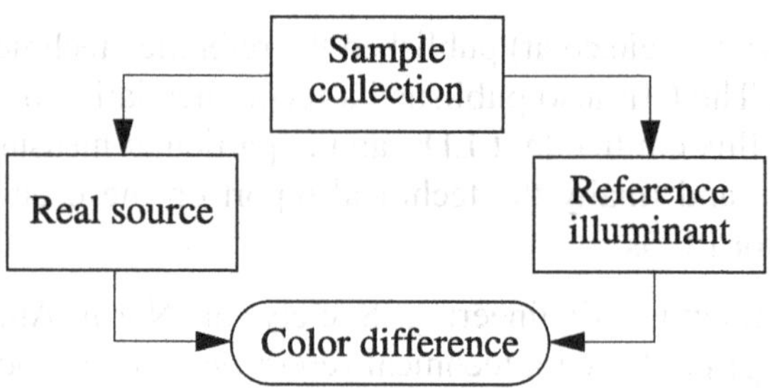

Figure 7.16. *Chosen schematic of the calculation of the CRI*

The base for the samples is given by colors chosen from the Munsell book. Eight colors of moderate saturation are used to calculate the average CRI R_a. Six additional colors, bright red, bright yellow, bright green, bright blue, the spectrum of carnation and the spectrum of foliage are used for the specific CRI R_i. The 14 colors are accurately defined by their spectral reflectance (Figure 7.17). It is fairly difficult to find real samples.

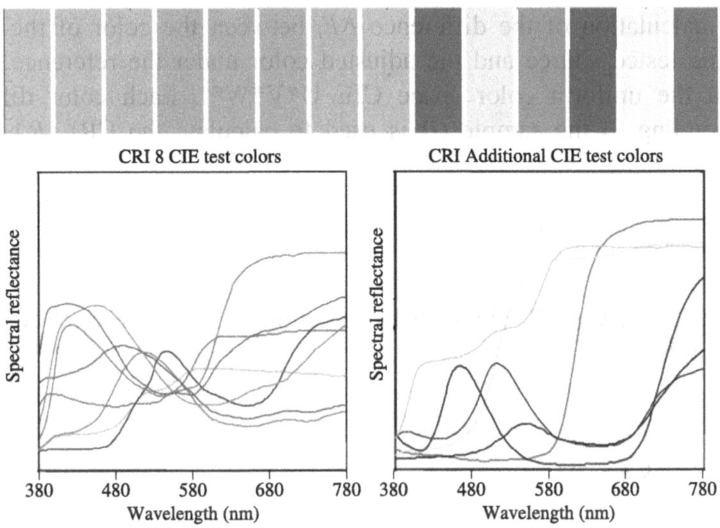

Figure 7.17. *Recommended test samples and reflectance factors for the CRI calculation (see color plate section)*

In principle, the reference light needs to have the same color temperature as the tested source and the best possible color rendering. The blackbody is chosen below 5,000 K and the appropriate daylight above 5,000 K (Figure 7.18). Because the reference light practically never has the same color as the tested source, an adjustment of its chromaticity is done to match its color with that of the tested source.

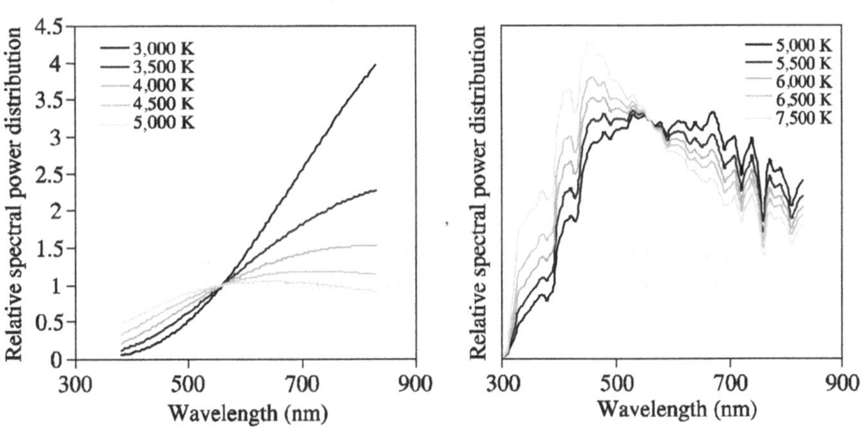

Figure 7.18. *Examples of relative spectral power distribution of the reference sources used for the color rendering calculation. Left: blackbody. Right: daylight*

The calculation of the difference ΔE_i between the color of the sample under the tested source and the adjusted color under the reference light is done in the uniform color space CIE U*V*W*[2]. Each color difference corresponding to the sample (i) is used to calculate the CRI (R_i) (Figure 7.19):

$$R_i = 100 - 4.6\Delta E_i \quad [7.11]$$

The average CRI R_a obtained by averaging the indexes (R_i) for all the eight samples is:

$$R_a = \frac{1}{8}\sum_{i=1}^{8} R_i \quad [7.12]$$

Equation 7.11 includes two constant factors chosen so that the average CRI is equal to 100 for the reference source, and to 50 for a traditional poor quality fluorescent lamp (Table 7.5).

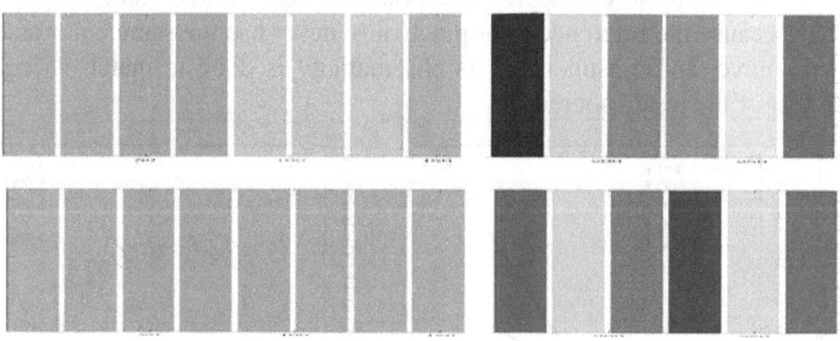

Figure 7.19. *Example of color differences obtained with a tested source and the reference source, after the CRI calculation (see color plate section)*

2. Note that this space, first mentioned in 1964 whereas other spaces are more recent, was in use at the time of the creation of the CRI. The *u*, *v* chromaticity diagram used in 1964 would nowadays be equivalent to a *u*', 2*v*'/3 diagram (see Figure 7.15).

Color rendering	CRI
Very good	≥ 90
Good quality	≥ 80
Poor	≥ 60

Table 7.5. *CRI table (AFE, 2006)*

7.4.2. *Calculation details*

The details of calculation are given in the CIE publication 13.3, with a floppy disk to run the calculation [CIE 95, SEV 96, SCH 07]:

– the input data are the values of the spectral reflectance of the samples and the relative spectral power distribution of the real source;

– the calculation of the color temperature of the real source dictates the choice of the reference source;

– if a shift of color remains between the real source and the reference source, it is corrected after the tristimulus values;

– the output values are the CRIs (Figure 7.20).

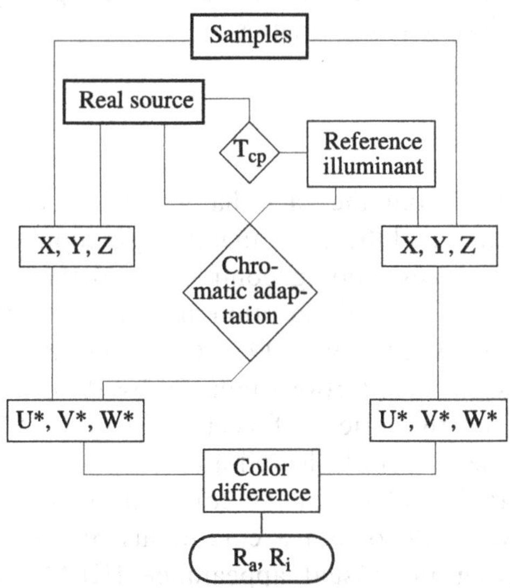

Figure 7.20. *Schematic of the routine for the CRI calculation, following the CIE*

7.4.3. *Update of the CIE position to take the observer's judgment into account*

Taking up some ancient questions, some recent experiments have shown the gap between the visual appreciation of the light emitted by LED sources and the CRI of the CIE. Attempts at modernization of the official formula such as choosing better targeted and commercially available samples, simplifying the chromatic adjustment or updating the color difference formula have not brought the improvements that were hoped for (see section 7.5.1.1). In 2007, the CIE acknowledged that the CRI was not suitable to predict the sorting of the color rendering for light sources when LEDs and traditional sources are compared [CIE 07b].

7.5. *Works on quality of light from LEDs*

The uprising of light emitting diodes in the markets of architecture lighting, scenic or valorization lighting, home, houses, shops or offices has been the subject of great interest and raises the question of quality of light in a greater sense. The word "quality" has many interpretations, the point of view and questioning of the end user being largely brought forward. Many studies have proposed to improve the description of color rendering. Some have a theoretical and modeling approach, whereas others instead emphasize real activities.

7.5.1. *Models*

Many teams start from the fact that color rendering is a rich and complex phenomenon, and therefore that a single number such as the CRI is too simple. All suggest adding complexity to the formula and more accuracy to the phenomenon description and to its effects on the various parts of color range. Already by 1995, R&D engineers of Philips noted that all the color surface distortions induced by fluorescent tube lighting could be represented by a field of vectors in the CIELAB chromacity plane, and that these vectors followed a bimodal deformation law of the chromatic circle. If the values for the color differences were normalized by the chroma, we could obtain a continuous bimodal function of the tonality, representing the visual appearance [KEM 95]. For example, under a given fluorescent tube, the colors shift towards violet or towards green. Depending on the light source, the distortions can therefore be more or less pronounced (the vectors can be more or less long) and deviate in the different regions of the chromatic circle. The formalism of a vector field can be applied to any source. The actual proposal of the same authors consists of several levels of description, from the more detailed, describing the distortions for optimised sample collection, to a simplified version giving the average spreading of the saturation losses of the colors [VAN 07].

7.5.1.1. *Attempt to update the CRI calculation*

Many clues were proposed by a technical comittee of the CIE (TC 1-33, Chairman's report, CIE, 1999).

The proposed index is noted R_{96a}. Its calculation was based on:

– the test colors of the Macbeth Color Checker,

– only six reference sources[3]: D65, D50, blackbody at 4,200 K, 3,450 K, 2,950 K and 2,700 K,

– the chromatic adaptation formula from the CIE in 1994, to bring the operations linked to the test source and the reference source to their results under the source D65,

– calculation of the color differences in the CIELAB space.

7.5.1.2. *A new index: Color Quality Scale (CQS)*

At the National Institute of Standards and Technology (NIST), Davis and Ohno have conducted a systematic modeling of the color of many surfaces exposed to many different light sources: existing sources, LED samples and theoretical spectra. Comparing the calculated CRI and the deviation calculations in CIELAB to visual observations, the CRI was shown to inaccurately predict the visual observation when exposed by LEDs. Precisely, the occurring of peaks in the LEDs spectra severely penalizes the CRI, although it contributes to an increase of the color level

3. Note that choice of the reference source was largely debated. The CRI calculation method recommended by the CIE suggests relating this choice to the color temperature of the tested source, which ends up with many reference sources. A reduced panel of reference sources could simplify the use, by limiting a few references for everyday life: home atmosphere, office, brighter atmosphere, etc.

often required for commercial applications. The choice of test samples is often critical for modeling the color rendering. Some can obtain excellent values, whereas others obtain poor indexes. Following this work, the authors have developed a single index, the CQS, which would reflect more accurately than the CRI the impression of an observer [OHN 05, DAV 05].

Considering that the color rendering is not limited to the color rendering fidelity, the authors (2005) recommended three topics:

– a new choice of test samples with fewer uncertainties,

– a more balanced estimation of the tonality and saturation levels,

– a calculation of the chromatic adaptation no longer related to the color temperature.

7.5.1.2.1. The choice of test samples

The test samples (only eight nowadays) do not properly reflect the color rendering. They are not saturated enough, even though it is well known that the color rendering can be degraded for saturated colors, especially if the source has an uneven spectral power distribution, as LEDs do.

To calculate the CQS, the new collection includes 15 saturated samples, well distributed along the chromatic circle and representing all existing colors. They are taken from the Munsell book and are commercially available. Their luminance factors are well known.

7.5.1.2.2. The balanced estimation of tonality and saturation levels

Today, all deviation of one of the attributes of the color sensation, whatever it is, is penalizing. The saturation gain which can be observed with diodes is often a desired effect for lighting because it improves the visibility and brightness of the scene. This saturation gain is ignored in the calculation of the color shifts in order not to penalize the source.

To obtain the CQS, the calculations are performed in uniform CIELAB space, which is often used and recommended.

7.5.1.2.3. The end of the reference to color temperature

Today the reference source is, by principle, optimal for color rendering. This is, however, not true for all color temperatures. Extreme color temperatures, as low as 3,000 K or as high as 7,500 K, do not give a rendering as satisfying as natural daylight. The reference to color temperature should be tempered.

In order to prevent this default, the authors propose to temporarily attribute to each color temperature a degrading factor calculated on the volume covered by the 15 samples in the CIELAB space. Considering that this volume is optimum at 6,500 K, the factor is equal to 1 for this color temperature. It is generally lower than 1 for the other color temperatures, except a few ones where it will be arbitrarily rectified to 1.

7.5.1.2.4. The CQS value

Finally, the average color variation is obtained by calculating the mean square of the shifts for the 15 samples, which gives more weight to the large shifts. The normalization factor of 4.6 taken from the CRI calculation has been temporarily adjusted to 3.01 to find the usual values of 50 and 100 for the CRI.

To avoid obtaining negative values of the new index, which was the case for the CRI of unacceptable lamps, the authors consider a non-linear transformation which would only affect values below 20, making them all positive.

For the authors, the new index is a quality index, the CQS.

7.5.1.2.5. Numerical simulations

The same authors have developed a program to generate the typical energy spectral distribution of a diode with a Gaussian profile, with a maximum at λ_{max} and a full width at half-maximum $\Delta\lambda$, and an adjustment program of the intensities of the RGB combination to obtain a given chromaticity [OHN 05].

The authors first mention a RGB combination with a CRI of $R_a = 63$, which gives a CQS of 80, thanks to the saturation gain of some of the samples. They then compare two RGB combinations with CRI $R_a = 80$, which are mostly different due to the choice of the red LED. The first combination ($\lambda_{max} = 603$ nm) induces large shifts on some of the samples which lowers the CQS to 73. The second combination ($\lambda_{max} = 613$ nm) increases the saturation of a few samples which increases the CQS to 85.

The simulation of the colors obtained on a screen seems to confirm the predictions of the CQS. Studies are still ongoing to improve the proposal.

Table 7.6 gives the CQS values for the lights whose relative spectral power distributions are shown in Figures 7.13 and 7.14.

	RGB	**WWRGBA**
$\boldsymbol{R_a}$	22	97
CQS	52	95

Table 7.6. *CRI and CQS values for a combination of red green and blue diodes (RGB) and a combination of warm white with one phosphor, cool white, red, green, blue and amber diodes (WWRGBA)*

7.5.2. *Color simulations*

Visualization techniques offer a great versatility to simulate the effect of a light source on a scene. From the recording of the spectral power distribution of all the elements of a scene, Nakano reproduced the scene just as it would appear if exposed to LEDs, a fluorescent tube or an incandescent lamp. Asking the observers to evaluate their visual impression on semantic scales, such as faithful/unfaithful, warm/cool, saturated/unsaturated, nice/ugly, etc., he pointed out two factors by analyzing the main components. The first factor is linked to the perception of the colorfulness, for 70%, the second to fidelity, for 16%, the latter being well correlated to color rendering [NAK 04].

The simulation in a visual display of a scene shown by several different lights, from its spectral characteristics, offers a versatility of use which makes it possible to rapidly grade different types of lighting. Then again, the correlation between the perception of the observers and the calculated CRI is not relevant [SCH 07].

7.5.3. *Experimental validations*

The rigorous experiments of real situations make it necessary to build adequate light cabins or specially equipped test laboratories with prototypes of LED-based lighting systems (Figure 7.21), in order to measure the spectral power distribution of the light emitted *in situ* and to calibrate all the required equipment. The observers, 10 to 40 people depending on the experiments, are generally young students, whose sight and color vision are normal. The tasks to achieve and the objects or materials tested depend on the assigned goals.

Figure 7.21. *Experimental handling of colored objects in a cabin exposed to LEDs*

7.5.3.1. *Discrimination*

Considering that the CRI accurately describes the color fidelity of objects exposed to natural daylight, the team of Viénot at the "Muséum National d'Histoire Naturelle" (MNHN) attempted to refine the quality judgments and focused on the discrimination accuracy and to the lack of apparent distortion of the colors [MAH 09]. Sight has a complex functioning and the judgment of the observers on the quality of light overcomes the global judgment on fidelity.

In practice, the authors have built a light cabin to install various combinations of LEDs: RGB, RGBA (red, green, blue, amber) combinations, white LEDs corrected by a few red LEDs, and also a large combination of many white and colored LEDs. The reference source was a tungsten halogen source with dichroic reflector, emitting at 4,000 K a light close to blackbody radiation or to direct sunlight.

40 observers were asked to sort colored samples with very similar colors and forming all together a chromatic circle (Figure 7.22). The test was purposely fairly difficult so that the subjects would make a few sorting mistakes. The assumption was that, if the subjects often make a sorting mistake with one source, it is the source which is to blame. The results showed that mistakes are twice as likely under RBG and RGBA sources than for other LED combinations or under the reference source [VIE 07, SZA 07]. The mistakes are mainly located around the reddish and greenish tones.

Figure 7.22. *Color discrimination test [VIE 07] (see color plate section)*

7.5.3.2. *Appearance*

In a peculiar way, the subjects noticed that under RGB and RGBA sources, which were the main sources of errors, the colors of the samples seemed brighter than under the reference source. The same authors then built up a second experiment [VIE 07, VIE 08]. 20 subjects were asked to evaluate on a scale of 0 to 10 the colorfulness of colored samples, also forming a chromatic circle. The results show that under RGB and RGBA, the reddish and blue greenish samples gained in apparent coloration.

7.5.3.3. *Preference*

At the LASH of the Ecole Nationale des Travaux Publics de l'Etat (ENTPE), the team of Fontoynont and Boissard compared lighting systems based on LEDs, fluorescent tubes and incandescent lamps. The observers were asked to reproduce the perception of fluorescent or incandescent lights with white, red, cyan and amber LEDs. A color target and the reproduction of painting had been installed in the light cabins. The subjects had to adjust the intensities of the various families of LEDs and it was suggested to start with warm white or cool white LEDs. The authors drew two conclusions. First, based on the appearance of the colored surfaces, the observers correctly succeeded in the LED adjustment trial. Then, three-quarters of the observers preferred the LED lighting to the incandescent light, and almost half of them preferred the LED lighting to the fluorescent light. In this configuration it seems that LEDs are well accepted and often preferred, which does not show up from a traditional calculation of the CRI [BOI 07].

7.5.3.4. *Harmony*

At the University of Pannonia, in Hungary, the team of Schanda and Bodrogi evaluated various aspects of the color rendering of LEDs. The color rendering of warm white LEDs (CRI = 98, 85 or 72) and RGB combination (CRI = 38) was compared to that of a fluorescent lamp (CRI = 85) and to a halogen lamp, directly on a color map. The CIECAM02 color appearance model predicts the results correctly [SAN 04]. The sorting task consisted of arranging samples with slightly different colors; white LEDs with good CRI gave the best results, while the RGB combination gave the worst. The harmony evaluation leaned on the judgment of academic combinations: pairs or triplets of colors taken from the Munsell book. The most appreciated sources were the white LEDs; the least appreciated were the RGB LED combinations. Finally, the appearance of common objects was also preferred under white LED exposure rather than under RBG combinations [SZA 07].

One of the main properties of the visual perception is the color consistency which comes from the capability of the visual system to disregard the variations of the lighting spectrum, which drastically change the spectrum and therefore the color radiated from the objects in the scene, to attribute them a color of its own. The color stability guarantees the correct identification of the objects in the scene. The colors are generally sorted into 11 categories: red, green, yellow, blue, orange, pink, violet, brown, white, black and gray. The classification in color categories of 292 color samples exposed to fluorescent light with good CRI (T_{cp}, R_a = 6,420 K, 96) and under four white LED light systems (blue LED plus yellow phosphor) with different color temperatures and CRI (T_{cp}, R_a = 7,500 K, 84; 5,100 K, 79; 4,000 K, 80; 2,800 K, 75) shows that the color rendering of these LEDs could be satisfying considering that their emitted light enables the correct classification of the samples [YAG 04]. A good overlap of the color categories improves the description of the color rendering.

7.5.3.5. *Comfort*

In addition to the color rendering, a few elements of comfort have to be taken into account by the light makers. In direct vision, the risk of glare can be serious with LEDs (see section 7.2.2.3). Component spacing [TAK 07] and light system reflectors can solve this problem.

In practice, the light manufacturers only characterize the sources with the color of light and CRI. However, the sensitivity of human beings is also affected by the circadian rhythm, the spectral sensitivity at sunset, the infrared radiation and the natural fluorescence of the materials. It is not impossible that our well-being is found when respecting these natural balances.

7.5.4. *Conclusion on the complexity of visual judgment*

We therefore reach a paradox. With some RGB or RGBA LED combinations, the apparent coloration of the colored surfaces of a scene can be improved, precisely where the fine color discrimination is lowered. We can conclude that for a proper color rendering of the surroundings and for the tone differentiation, these RGB or RGBA combinations should be avoided. The white diodes generate a much more satisfying light.

The visual experiments conducted with LEDs show that colorfulness perception can be as important as the color fidelity perception, and that colorfulness perception is lacking from the actual CRI definition.

Let us keep in mind that, in the visual system, the information treatment is both parallel and organized. The details and the color differences are treated locally in the retina. The integration of the color and the appearance takes place in the brain and includes all the visual fields. Ultimately, it is the observer's judgment which should dictate the quality grade.

7.6. Applications of LEDs to lighting

Lighting is usually achieved with discharge lamps outdoors (high pressure sodium or metal iodium), fluorescent tubes indoors and incandescent lamps or compact fluorescent lamps at home. These lamps radiate large quantities of light in all directions, and the light systems are designed to redirect the flux for the application needs. The performances reached by inside lighting lamps are indicated in Table 7.7. Only a few fluorescent tubes functioning on electronic ballasts with high luminous efficiency are compliant with the thermal regulations and the lighting renovation rules for building and constructions, which make it necessary to install electrical powers less than 2.5 or 2.8 W/m^2 for 100 lux in the working area.

	Fluorescent tubes	**Compact fluorescent lamps**	**Incandescent lamps**
Unit flux (lm)	1,350 to 7,000	1,000 to 1,800	
Luminous efficiency (lm/W)	80 to 104	33 to 64	10 to 20
Color temperature (K)	2,700 to 6,500	2,700, 4,000, 6,500	2,700 to 3,000
CRI	80 to 93	85	100
Lifetime (h)	10,000 to 16,000	10,000 to 16,000	1,000 to 2,000

Table 7.7. *Performances of common lighting lamps [documents from AFE]*

With today's technology, and despite the manufacturers' announcements, the available luminous flux of LED-based technology is still low, and the use of LEDs is limited to confined environments. Even if LEDs cannot replace classical sources yet, they have many advantages that allow them to answer many specific needs [TAG 06]. These advantages are presented in Chapter 1.

7.7. Conclusion: advantages, precautions and perspectives

The quality of artificial light is assessed with respect to natural daylight. For lighting applications, not only has the light color to be managed, but also the color rendering of the objects exposed to this light. Colorimetry offers many tools to characterize the color rendering by the calculation of the CRI of the sources.

It appears, when observing real scenes exposed by LEDs, that quality of light appreciation is a complex visual phenomenon and cannot be accurately described only by the CRI. However this question evolves in the future, spectrum management will be in any case the key of success for LED-based lighting.

The research activity around the quality of light emitted by LEDs is intense and it is very likely that new recommendations will arise from the CIE when new light sources appear in everyday life.

7.8. Acknowledgements

My acknowledgements go to all my colleagues, in particular Jean-Jacques Ezrati, Elodie Mahler, Albane Rambaud, Bernard Duval, Christian Eugène, Alain Bricoune and Frantz Dennery.

7.9. Bibliography

[AFE 06] ASSOCIATION FRANÇAISE DE L'ECLAIRAGE, "Recommandations relatives à l'éclairage intérieur des lieux de travail", 2006.

[ANS 07] ANSI, "Specifications for the chromaticity of solid state lighting products", *ANSI C78.377*, 2007.

[BOI 07] BOISSARD S., "Optimisation of mixing of LED-based light for object presentation. Comparison with traditional light sources", *CIE 26th session*, Beijing, China, p. D1:30-33, 2007.

[CIE 87] CIE, "International lighting vocabulary", CIE 17.4, 1987.

[CIE 95] CIE, "Method of measuring and specifying color rendering properties of light source", CIE 13.3, 1995.

[CIE 02a] CIE, "Glare from small, large and complex sources", CIE 147, 2002.

[CIE 02b] CIE, "CIE equations for disability glare", CIE 146, 2002.

[CIE 04] CIE, "Colorimetry, 3rd edition", CIE 015, 2004.

[CIE 04] CIE, "A color appearance model for color management systems: CIECAM02", CIE 159, 2004.

[CIE 07a] CIE, "Measurement of LEDs", CIE 127, 2007.

[CIE 07b] CIE, "Color Rendering of white LED light sources", CIE 177, 2007.

[DAV 05] DAVIS W., OHNO Y., "Toward an improved color rendering metric", *Proceedings of the SPIE*, vol. 5941, p. 1G1-6, 2005.

[IES 07] IESNA, "Technical memorandum on light emitting diode (LED) sources and systems", IESNA TM-16-05.

[KEM 95] VAN KEMENADE J.T.C., VAN DER BURGT P.J.M., "Towards a user oriented description of color rendition of light sources", *Proc 23rd Session*, New Delhi, CIE 119, p. 43-46, 1995.

[LI 06] LI, C., LUO, M.R., "Quality of LED based daylight simulators", *CGIV 2006*, 247-250.

[LUM 07] LUMILEDS, "Understanding power LEDs lifetime analysis", White Paper, http://www.lumileds.com/docs/docs.cfm?docType=16, 05/22/2007.

[MAH 09] MAHLER E., EZRATI J.-J., VIÉNOT F., "Testing LED lighting for color discrimination and color rendering", *Color Research & Application*, vol. 34, p. 8-17, 2009.

[NAK 04] NAKANO Y., TAHARA H., SUEHARA K., KOHDA J., YANO T., "Color rendering simulator using multispectral camera and its application to evaluating white LEDs light sources", *CIE x026,* p. 132-133, 2004.

[OHN 05] OHNO Y., "Spectral design considerations for white LEDs color rendering", *Optical Engineering*, vol. 44, p. 111302, 2005.

[SAN 04] SÁNDOR N., BODROGI P., CSUTI P., KRÁNICZ B., SCHANDA J., "Direct visual assessment of color rendering", *Proc. CIE 25th Session*, San Diego, US, p. D1: 42-45, 2003.

[SCH 07] SCHANDA J., *Colorimetry, Understanding the CIE System*, Wiley, Hoboken, 2007.

[SEV 96] SEVE R., *Physique de la couleur. De l'apparence colorée à la technique colorimétrique*, Paris, Masson, 1996.

[SPE 06] SPEIER I., SALSBURY M., "Color temperature tunable white light LED system", Sixth international conference on solid state lighting, *Proc. of SPIE*, Vol.63371F, p. 1-12, 2006.

[SZA 07] SZABÓ F., SCHANDA J., BODROGI P., RADKOV E., "A comparative study of new solid state light sources", *Proc. CIE 26th Session*, p. D1:18-21, 2007.

[TAG 06] TAGUCHI T., "Developing white LED lighting systems and its technological roadmap in Japan", *Journal of Light and Visual Environment*, vol. 30, n°3, p. 65-70, 2006.

[TAK 07] TAKAHASHI H., IRIKURA T., MORIYAMA T., TODA M., IWAMOTO M., "Discomfort glare and annoyance caused by white LED lamp", *CIE 26th Session*, Beijing, China, p. D1 : 80-83, 2007.

[VAN 07] VAN DER BURGT P.J.M., VAN KEMENADE J.T.C., "About color rendition of light sources. The balance between silplicity and accuracy", *CIE 26th Session*, Beijing, China, p. D1:26-29, 2007.

[VIE 07] VIÉNOT F., EZRATI J.-J., BOUST C., MAHLER E., "Grading LED illumination: from color rendering indices to specific light quality indices", *CIE 26th session*, Beijing, China, p. D1:22-25, 2007.

[VIE 08] VIÉNOT F., MAHLER E., EZRATI J.-J., BOUST C., RAMBAUD A., BRICOUNE A., "Color appearance under LED illumination: the visual judgment of observers", *Journal of Light and Visual Environment*, vol. 32, p. 208-213, 2008.

[WYS 82] Wyszecki G., Stiles W.S., *Color Science Concepts and Methods, Quantitative Data and Formulae*, Wiley, New York, 1982.

[YAG 04] YAGUCHI H., ENDOW N., MORIYAMA T., SHIOIRI S., "Categorical color rendering of LED light sources", *CIE x026*, p. 20-23, 2004.

Chapter 8

OLED Technology

8.1. Introduction

8.1.1. *Organic materials: a history*

It was in 1963 that a few scientists first became interested in some kinds of organic materials and in their optoelectronic activity. This work dealt with anthracene crystals (POP 63) electroluminescence. It was the first optoelectronic attempt on an organic material. However, the work on anthracen crystals has remained at a fundamental research level, requiring a much too high driving voltage (>300 V) and not being reliable enough to be exploited for commercial electronic devices.

At the beginning of the 1980s, anthracene came back into fashion. Scientists [ROB 82] managed to noticeably reduce the driving voltage of anthracene-based devices. To do this, they used thin layer techniques to spread this molecular crystal into thin amorphous layers. Nonetheless, device efficiency was low, too low to consider the commercial exploitation of a system. It will take at least 10 years for the organic revolution to finally take place in optoelectronics.

Chapter written by Tony MAINDRON and David VAUFREY.

8.1.2. *Birth of the first OLED device*

In 1987, a Kodak Team, initially dealing with solar photoreceptors, worked on an electroluminescent device with materials coming from xerography [TAN 87]. Indeed, they worked out the first organic electroluminescent diode with two sublimed juxtaposed thin layers. The stack, almost the same as that in use today, was as follows: ITO/diamine/AlQ3/MgAg. This was the birth of the OLED (Organic Light-Emitting Diode) acronym, now frequently used to refer to electroluminescent organic diodes formed of small molecules. In these early devices, the two different layers have very distinct electrical properties: the first layer, made of a diamine (TPD standing for (N,N'-diphenyl-N,N'-di-(m-tolyl)-p-benzidine)) is a hole carrier and the second, the AlQ3 (standing for tris(8-hydroxyquinoline) aluminum) layer, is an electron carrier and the luminescent center.

In 1990, a British team achieved an electroluminescent diode similar to the one described above yet with a semiconductor polymer: poly(p-phenylene vinylene) [BUR 90]. This material simultaneously worked as an electric charge bipolar carrier and a luminescent center. The PLED (*Polymeric Light-Emitting Diode*) acronym was born. It is now commonly used and more specifically designates electroluminescent devices made of semiconductor polymers instead of those made of small molecules and hence designed by the above OLED acronym. Nevertheless, the OLED acronym remains the preferred term by default to describe this technology from a general point of view.

8.2. Electroluminescent diodes

In an OLED device, charges are electrically generated by applying a potential to the structure electrodes. Hence, considering structures made either of sublimed small molecules or polymers, a process leading to light extraction by luminescence may be described as follows:

– negative (resp. positive) charge injection in the electron (resp. hole) carrying a thin layer by application of an electric potential to the respective electrodes;

– electrical charge transport across the above-mentioned layers thanks to respective radical ions (oxido-reduction reactions);

– Coulombian capture of two opposite charges inside the luminescent material;

– formation of an exciton on a luminescent material molecule. The excess of energy coming from the recombination enables the excitation of a molecule;

– electromagnetic radiation due to the return of the molecule to its fundamental state. The emitted photon wavelength then depends on the material. For example, with AlQ3 green light, around 520 nm is emitted.

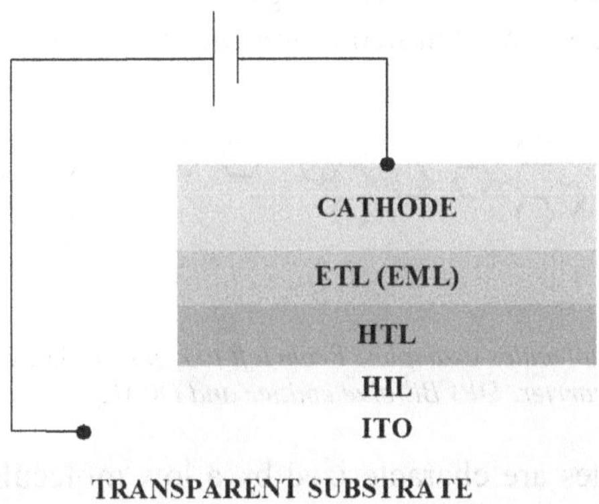

Figure 8.1. *OLED typical stack*

Moreover, the manufacturing of organic flat screens and lighting sources depends on the ability to make the three primary colors, red, green and blue, which are the basis of these devices. These colors must be saturated and remain so, at least during the announced lifetime of the component. This implies a good temporal stability of the pigments and hence a good electrochemical stability. White organic emission is usually performed by an OLED exhibiting two or three primary pigments in the same structure.

Today the main challenge is the stability of OLED integrated materials. These devices must achieve a significant lifetime in order to compete with inorganic semiconductor devices (such as III-V compounds).

An OLED long lifetime may only be reached by selecting high purity materials, careful device processing under controlled atmosphere and protecting (through encapsulation).

The final goal is to prevent chromophores (color source structural units inside the material) from decay due to the combined effect of light, oxygen or moisture. The OLED may also quickly fade if protection against atmosphere attacks is not taken.

8.2.1. *Organic semiconductor categories*

8.2.1.1. *Small molecules*

Small molecule OLED technology has been developed by Eastman Kodak. Figure 8.2 shows detailed representations of some small molecules.

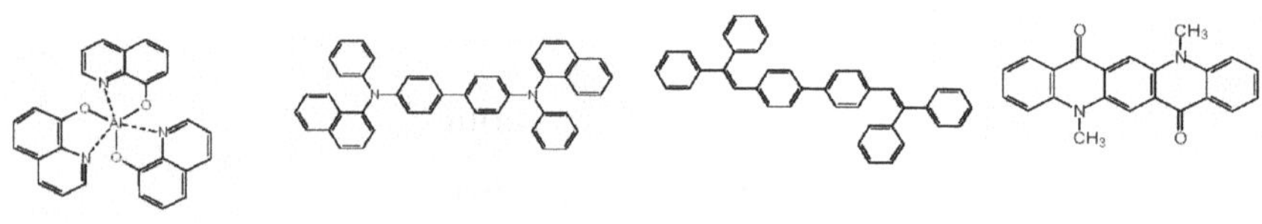

Figure 8.2. *Small molecules examples. From left to right: AlQ3*[1]*, green emitter, NPB*[2] *hole carrier, DPVBi*[3] *blue emitter and DCM2*[4] *red emitter*

Small molecules are characterized by a low molecular mass. Optical and electrical properties can be optimized by judicious choice of the groups on the periphery or ligands (for organometallics). Thin layer deposition is obtained using simple methods derived from evaporation.

8.2.1.2. *Polymers*

By comparison, polymers have long molecular chains and hence a greater molecular mass. Several forms may be encountered: homopolymers, copolymers and dendrimers. Homopolymers are composed of a monomer chain, whereas copolymers result from the association of two or more different monomer repeated several times. Electrical and optical properties then depend on the type of monomers and the periphery groups, as shown in Figure 8.3 describing PPV and polyfluorene variants.

---

1. AlQ3: tris-(8-hydroxyquinoline)aluminium.
2. NPB: N,N'-Bis(naphtalen-1-yl)-N,N'-bis(phenyl)benzidine.
3. DPVBi: 4,4 '-bis(2,2 ' diphenyl vinyl)-1,1 '-biphenyl.
4. DCM2: 2-methyl-6-[2,3,6,7-tetrahydro-1H,5H-benzo[ij]quinolizin-9-yl ethenyl]-4H-pyran-4-ylidene] propane-dinitrile.

Electronic transport in polymers is similar to that of semiconductors since charges can delocate themselves on HOMO[5] and LUMO[6] levels along the polymer. Intramolecular conductivity is then mentioned. Nevertheless, like small molecules, charges must jump from one molecule to the other (*hopping* model); polymers thus have a second type of conductivity, weaker than the first one described and called intermolecular conductivity.

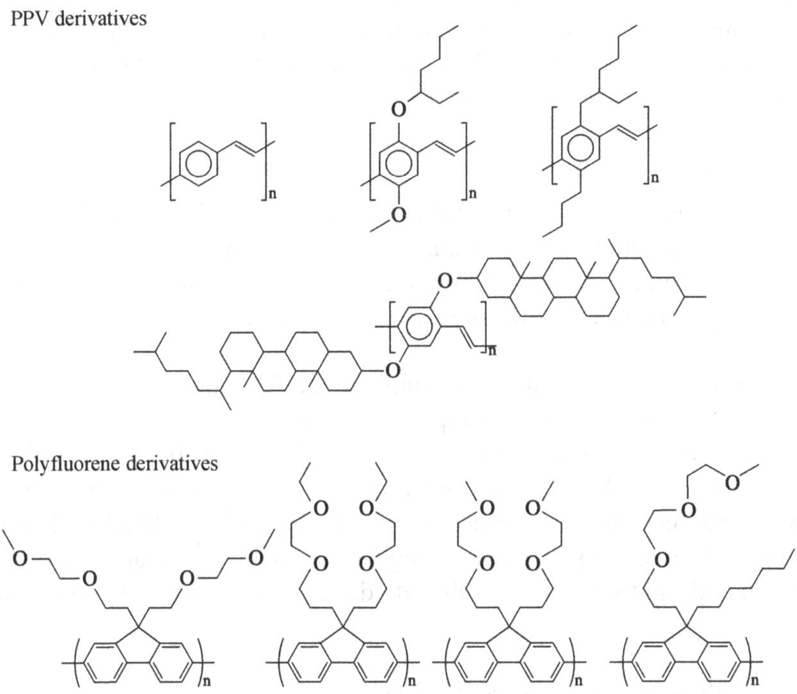

Figure 8.3. *Some polymer examples and the evolution of their periphery groups added to improve their electro-optical properties*

Polymers are wet processed by spin coating or another variant of the wet deposition process. This kind of process may also be applied to some types of small molecules.

5. HOMO: Highest Occupied Molecular Orbital.
6. LUMO: Lowest Unoccupied Molecular Orbital.

8.2.2. *Deposition technique description*

For electroluminescent diode fabrication, a thin layer of small molecules and polymers must be deposited. Films, whose thickness does not exceed 100 nm, must be homogenous and free from pinholes. The ability to be well deposited in thin layer is hence the *sine qua non* condition for these materials. For small conjugate molecules, it implies the possibility of secondary vacuum evaporation without decomposition during heating (which means that these fragile materials must have a high vapor pressure). For polymers, they must be soluble in standard solvents to be deposited by spin coating or the ink jet method.

8.2.2.1. *Evaporation of mall molecules*

This process under vacuum leads to organic thin layers with very few impurities. Impurities may contribute to an accelerated decay of devices or to a decrease in performance. However, the electron beam should be avoided because it generates a significant modification of material properties.

In addition, the thin layers of small molecules cannot be structured by classical photolithography techniques, so a stencil is frequently used. It is generally a metallic plate perforated with the desired shapes to be replicated on the substrate. This microstructuration method implies an alignment between the mask and the substrate inside the deposition chamber. It is also limiting for large substrate processing or for high resolution patterns. For OLED-based lighting sources, substrate dimensions will thus be the limiting factor.

8.2.2.2. *Polymer wet deposition process*

The main drawback of the wet deposition process is the integration of impurities inside the film due to the not-under vacuum process conditions. It then requires perfect control of solvent purification. This is not an easy task for large scale fabrication. That is why polymers are slightly behind small molecules from a commercialization perspective.

The most frequently used deposition technique for polymers is spin coating. This technique is very easy to implement and allows deposition on large substrates at a lower cost than small molecules. However, as for small molecules, there is no microstructuration technique for these thin layers.

To compensate for the lack of microstructuration techniques, ink jet technologies or screen process have been developed. The ink jet process uses a printer and polymer in solution. Printer nozzles are adapted to exactly fill a tank which is usually surrounded by a hydrophobic barrier to prevent contamination of adjoining pixels.

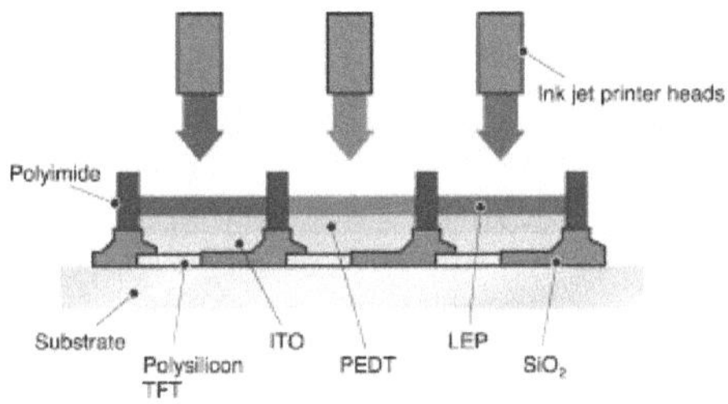

Figure 8.4. *OLED ink jet method for polymer deposition (here PEDT and LEP) (see color plate section)*

Liquid solution structuration is obviously a more economical way to produce OLED devices.

8.3. Organic semiconductors: theory

8.3.1. *Introduction to semiconductivity in organic chemistry*

A material semiconducting property is due to the presence of a forbidden bandgap in its electronic structure. It also applies to organic semiconductors.

8.3.1.1. *Band theory in organic chemistry*

To understand the type of bond and the electric properties from a chemical point of view, let us study the case of a small amount of metallic atoms coming into interaction. When two atoms get close enough, their orbitals interact. Two wave functions also interact: one of the resulting wave functions has a greater energy (called antibonding) and the other function has a lower energy (called bonding). Consequently, the interaction between n atoms will generate n very close energy levels, some of them being bonding, some antibonding (the whole set is called a band) (Figure 8.5).

In a metal, the band will be half-filled, the most stable energy levels being doubly filled whereas the remaining half – higher energy levels – are empty. This is only true for absolute zero observation. For other temperatures, many levels are partially filled due to Boltzmann distribution and energy level proximity, therefore the limit is blurred. The highest fully filled energy level is called the Fermi level.

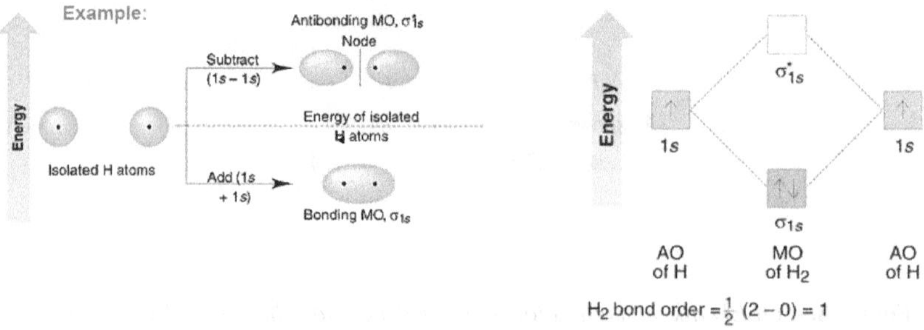

Figure 8.5. *Bond formation between two atoms and consequences on electronic levels*

8.3.1.2. *Differences compared to classical semiconductors*

A notable difference for organic solids compared to inorganic solids is that molecular entities are held together thanks to van der Waals interactions, whereas intramolecular bonds are stronger. As such, the interactions between molecules are weak and the free mean path is roughly equal to the intermolecular space. As a consequence, the band structure of organic materials is localized and not delocalized on the whole structure.

In summary, since interactions are weak, an organic solid structure and a single molecule have many characteristics in common. For a molecule, potential wells induced by each nucleus join together for the higher energies and form a large well.

Deep atomic orbital are localized in the nucleus potential well, whereas higher atomic orbitals interact to form delocalized atomic orbitals, but only on the molecule. Two levels may be distinguished: the HOMO (*Highest Occupied Molecular Orbital*) level and the LUMO (*Lowest Unoccupied Molecular Orbital*) level, which can both be compared to the valence and conduction band respectively for inorganic semiconductors, yet localized at the molecule level.

In organic solids, charge transport is incoherent. Every transport site is localized because disorder potential is much greater than the ratio of ordinated structure. Conductivities in organic solids are hence lower than in inorganic solids. From a chemical point of view, charge transport occurs through an oxido-reduction process between molecules or pieces of molecules.

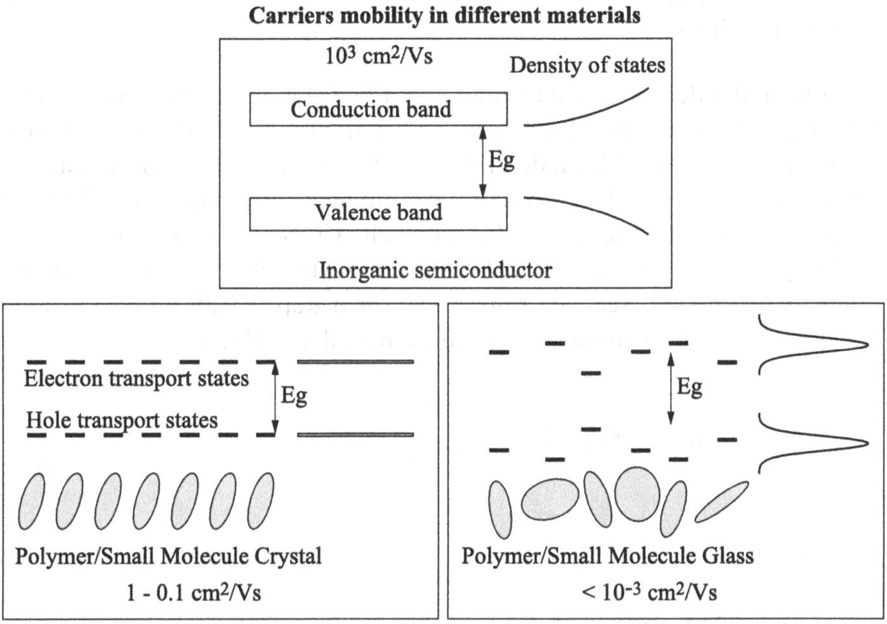

Figure 8.6. *Comparison between electronic band structures in organic semiconductors and electronic levels in organized organic semiconductors (bottom left) and unorganized organic semiconductors (bottom right)*

As described in Figure 8.6, amorphous organic solids have two densities of localized levels merged into Gaussian forms. This explains the frequent use of Brownian motion laws or Monte Carlo algorithms to model the transport of charges.

On the other hand, weak intermolecular bounds allow mechanical flexibility without penalizing defects formation. It is a great advantage for emissive device feasibility on flexible substrates. Consequently fabrication of conformable displays or lighting devices becomes possible.

8.3.2. *Electronic transport model in amorphous organic solids*

In the case of amorphous molecular solids (which are the core of OLED devices), electronic transport is of a localized kind on the individual molecule. Usually, localized jump from molecule to molecule (hopping model) formalism is applied to describe the transport of a charge. This model has been developed initially to describe transport in solids such as amorphous silicon.

This model describes a tunneling effect between molecules sustained by network vibration. This formalism was initially developed for classical semiconductors. It enabled a description of the movements of charge carriers inside unordered covalent crystals with numerous structural or chemical defects. It was first described by Conwell [CON 79] and then put into mathematical form by Miller and Abrahams [MIL 60]. The transition rate (number of tries per second) from a site, or a trap, i with an energy ε_i to another site j with an energy ε_j may be expressed as follows:

$$\nu_{ij} = \nu_0 \exp\left(-2\gamma\Delta R_{ij}\right)\exp\left[-\left(\frac{\varepsilon_j - \varepsilon_i}{kT}\right)\right] \qquad [8.1]$$

for $\varepsilon_j > \varepsilon_i$, and:

$$\nu_{ij} = \nu_0 \exp\left(-2\gamma\Delta R_{ij}\right) \qquad [8.2]$$

for $\varepsilon_j < \varepsilon_i$.

ν_0 is the phonon vibration frequency corrected by a probability of tunnelling effect for jumps towards lower energy states ($\varepsilon_j < \varepsilon_i$). This probability depends on the intersite distance (ΔR_{ij}), a γ term, representing the overlapping of the wave functions of states i and j. A Boltzmann distribution, which describes a phonon absorption probability, is added in the case of jumps towards higher energy ($\varepsilon_j > \varepsilon_i$).

If the hopping theory is applicable for all organic semiconductors, the formalism used to describe charge carriers mobility in these materials on the contrary depends on the nature and the structure of the material.

Polaron formalism [GIL 72] is usually applied to conducting polymers and molecular crystals, which have a narrow band structure. Disorder theory [BAS 81] describes best mobility in amorphous organic glasses.

A polaron is an electron (or a hole) trapped in a self-induced potential. In fact, time spent by the charge on the molecular site is greater than its hopping time. The interaction between the charge and the site is stronger because the electron stays on the site longer than the vibration period of the molecule nuclei (which is roughly 10^{-14} s). Thus, during all the time spent by the charge on the molecule, the molecule nuclei will change of equilibrium state (due to the electron presence) with a different polarization. The electron, first causing an induced polarization on the molecule, will then be subjected to this polarization. The electron energy will be reduced and it will hence be trapped in a self-induced potential. The result (charge + induced polarization) is the polaron. The polaron cannot move without getting over the potential barrier. It mainly proceeds by jumping. The potential barrier is smaller for an electron in an excited state than for an electron in its fundamental state, so the hopping process is a charge transport by the excited states. Hopping probability increases for excited states with a long lifetime.

The polaron model has a severe limitation and does not always allow a proper modeling of the observed behaviors. Indeed, it does not describe the dependence between mobility on the field and temperature, as shown experimentally in the case of amorphous molecular solids. Bässler disorder formalism is then preferred for this latter example.

This formalism considers that when there is no complete order, positions and energy of molecular sites are described by the random Gaussian distribution of localized states. Energetic distribution of the sites is due to permanent disorder caused by dipole-dipole fluctuations between molecules that interact with the additional electron of a nearby molecule. Energetic and spatial distributions of states are assumed to have Gaussian shapes because each one depends on a great number of parameters, each having microscopic variations.

Transport can then happen towards energetic sites with lower or higher energy, in the intrinsic states density (transport band), in the forbidden band (traps) and from transport band towards the forbidden band. Jumps towards higher energical states require electrical field support if the thermal activation energy is not enough.

Since the presence of an electric field alters the energetic distribution of the sites and reduces the average bandgap oriented in its direction, it can be said that charge mobility in amorphous organic solids strongly depends on the field. Knowing that charge mobility in these materials also depends on temperature, mobility μ(E,T) for amorphous organic solids can be describe by the following typical mathematical expression:

$$\mu(E,T) = \mu_0 \exp\left[-\left(\frac{2\sigma}{3kT}\right)^2\right] \exp\left\{C\left[\left(\frac{\sigma}{kT}\right)^2 - \Sigma^2\right]\sqrt{E}\right\} \qquad [8.3]$$

This empirical formula well describes experimental observations made on amorphous organic solids. It was first used in 1970 to describe experimental data on mobility in polyvinyl carbazole thin films.

Here are the bases of the formalism:

– energetic sites and distances are marked by Gaussian distributions with half height width called σ and Σ respectively;

– *hopping* is described by the previously mentioned Millers-Abraham formalism;

– electron phonon coupling is sufficiently low to neglect polaron formation;

– the process remains incoherent and each jump is independent from the others;

– material traps are shallow, that is, close to bands and hence there is a critical temperature for traps, Tc, below which transport becomes dispersive. The disorder model only applies for Tc<T<Tg[7];

– the model is only valid for large fields, typically $>10^5$ V/cm. It is the order of magnitude of the field encountered in OLEDs.

7. Tg: organic material glass transition temperature.

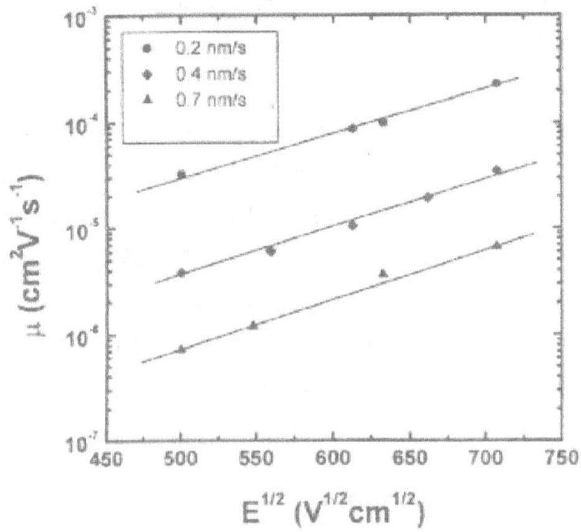

Figure 8.7. *Time-of-flight electron mobility measure in an amorphous layer of AlQ3 deposited at different evaporation rates [CHE 99]*

Other models have been elaborated [BOR 98]. However, disorder formalism describes best observations made on mobility in amorphous organic semiconductors, in particular the mobility logarithm dependence on the field like $SE^{1/2}$ and the mobility dependence on temperature like $-(2\sigma/3kT)^2$ for zero field. As can be seen in Figure 8.7, typical mobility values for this kind of semiconductor are between 10^{-6} V/cm^2/s and 10^{-3}V/cm^2/s.

8.4. OLED electrical characteristics

8.4.1. *Charge carriers injection models*

Since the beginning of the 1990s, fundamental research carried out on OLEDs has tried to explain which factors were limiting device efficiency. Two limiting factors have been successively proposed: injection limited transport of charge between electrodes towards respective organic layers, and bulk limited transport due to material properties.

Injection mechanisms have been borrowed by classical semiconductor theory. They are described by the tunnel effect model (Fowler-Nordheim model) and thermo-ionic emission (Richardson-Scottky model) in the case of Schottky metal-semiconductor contact. Proposed bulk limited transport mechanisms are: ohmic conduction and space charge limited current (SCLC), with or without traps.

Today, theories point out the bulk limited transport and hence the space charge formation-related phenomenons. Besides, even if mobility dependences on the field and the temperature (as described by disorder theory and following $S.E^{1/2}$ and $-(T_0/T)^2$ respectively) have been firstly omitted, it now seems that they must be integrated into the equations describing transport. Thus, it improves the models and hence the understanding of this peculiar kind of transport.

8.4.2. *Charge carriers transport models*

In organic semiconductors, it has already been said that charge mobilities are low. When charge carriers are injected from a metal towards an organic solid under an electric field, charges accumulate in the solid, close to the interface, leading to significant space charges. Let us consider, from a simplified point of view, the injection of electrons from a cathode to an electron carrier organic solid, thanks to an electric field. Injected negative charge carriers create a global charge |-Q| close to the electrode (indeed, in a simplified view, their "evacuation kinetic" is slow, due to their low mobility, compared to their "injection kinetic"). It then makes sense to believe that this charge space |-Q| will in turn have consequences on the subsequent electron injection and will notably modify it.

We will now study the formalism describing SCLC in these semiconductors. Mathematical expressions have been developed in the unipolar charge transport case. For multilayered OLEDs, that is, for bipolar transport, theories get more complicated. Today, there is still no unified theory that takes into account simultaneously bipolar current, injection limited transport and bulk limited transport.

Even if advanced theories tend to describe charge transport in OLEDs by bulk limited transport theory, it seems obvious that charge injection into the organic semiconductor has consequences on the transport inside the material.

As an example, Mark and Helfrich have described SCLC theory for organic crystals [MAR 62]. This theory assumes some similarity with classic band theory. In particular, it resorts to valence band and conduction band notions, to the energy gap and to the traps distributed in this gap. It does not take into account mobility dependence on the field and temperature.

In the case of an organic semiconductor without traps, where mobilities are low, Child's law describes current law in permanent regime:

$$J_{SCLC} = \frac{9}{8}\varepsilon\mu\frac{V_{appl}^{2}}{d^{3}} \qquad [8.4]$$

$\varepsilon=\varepsilon_0\varepsilon_r$ is the material permittivity, d represents organic thin layer thickness, V_{appl} is the applied potential and μ is the carrier's mobility.

At low voltage, when the density of intrinsic carriers (thermic carriers whose distribution is given by the Fermi-Dirac statistic) is considerably greater than the density of injected carriers, Ohm's law controls conduction [SAV 97]:

$$J_{\Omega} = n_0 q\mu\frac{V_{appl}}{d} \qquad [8.5]$$

For this kind of semiconductor, the current depends on material thickness in both formulae. It should be noted that n_0, which describes the intrinsic of free carriers, has vanished from the J_{SCLC} expression in [8.4]. In fact, this current law, and hence its deviation from Ohm's law (which dominates at low voltage), occurs at greater voltage, as soon as the injected charge carrier density is greater than n_0. On the contrary, the classical conductivity of PN junctions is driven by intrinsic or extrinsic carriers (coming from doping).

Holes transport is done on low energy levels; hence holes are insensible to traps coming from environmental factors such as oxygen. For example, in PPV, their transport is not subjected to trapping [BLO 96]. On the contrary, electrons are sensitive to this kind of close traps and almost always present a dispersive transport. In a charge mobility measure experiment, for example using time-of-flight measurement, obtaining the electrons transit time in a given material is very difficult because the transitional current plateau is not significant [BOR 98].

The most realistic description of trap distribution is the mathematical exponential description of traps located close to the conduction band of electrons [BLO 96]. In this case, the state density of traps $n_t(E)$ has a Boltzmann distribution profile ($E<E_c$):

$$n_t(E)=\left(\frac{N_t}{kT_t}\right)\exp\left(\frac{E-E_c}{kT_t}\right) \qquad [8.6]$$

T_t, E_c and N_t are the trap temperature, the conduction band energy level and the total trap density respectively. Space charges with limited current that include an exponenetial distribution of traps are more complicated to describe. The formula is given by Mark and Helfrich formula [MAR 62]:

$$J_{TFL}=N_C\mu q^{(1-m)}\left(\frac{m\varepsilon}{N_t(m+1)}\right)^m\left(\frac{2m+1}{m+1}\right)^{m+1}\left(\frac{V^{m+1}}{d^{2m+1}}\right) \qquad [8.7]$$

N_C here represents the conduction band state density, exponent m is defined by $m=T_t/T$ and must be strictly greater than 1. The greater m is, the slower the distribution exponential $n_t(E)$ decreases and the more traps there are.

Recent theories point out the necessity of take mobility dependence on the field into account. Ultimately, there is nothing surprising in that, since it is the same materials used before in xerography that are now the active core of OLEDs. Yet, disorder theory fitted very well with experimental observations made on photoreceptors used, for example, in photocopiers. Consequently, this theory should logically contribute to transport description in OLEDs.

Ioannidis *et al.* [IOA 98] demonstrated that this expression could be added to the description of a permanent ohmic current, without taking into account space charge currents. They proved that in AlQ3-based OLED diodes, with an Al/AlQ3/Al structure where electrons are the majority carriers, the current/tension graph profiles could be well described by:

$$J=n_i e\mu_0\exp\left(S\sqrt{E}\right)E \qquad [8.8]$$

In this case, the material's own intrinsic carriers are neglected compared to injected carriers ($n_0<<n_i$) because of the large energy gap,

and traps are assumed to be shallow like in disorder theory. μ_0 represents zero field mobility.

Simulation satisfactorily explains current profiles following $J \propto V^{l+1}$ without integrating space charges and related hypothesis. Furthermore, it describes perfectly well the dependence of devices on thickness, in particular the stability of the (V/d^2) ratio considering constant current and high exponent.

At the same time, other groups keep the SCLC theory valid and include the now compulsory mobility dependence on field. Campbell *et al.* develop a model for J-V graphs following the expression [SCO 00]:

$$J = \frac{9}{8}\varepsilon\mu_0 \exp\left(S\sqrt{E}\right)\frac{E^2}{d} \qquad [8.9]$$

The theoretical research Holy Grail for OLEDs is the correct description of electro-optic characteristics shown by devices. For the time being, the description is still vague.

At the beginning, injection, considered in a tunnel effect context, was assumed to be the limiting factor of OLED efficiency. Today, thermo-ionic injection is now considered to be this limiting factor [CRO 98] given that potential barriers are low at interfaces so that the current is limited by space charges instead of injection.

8.5. Different structure types of OLEDs

8.5.1. *Direct and inverted diodes*

Beginning from the substrate, the layer stacks of OLEDs can either start with an anode or a cathode. They are then called a direct diode or an inverted diode respectively. The choice of one of these two stacks has consequences on OLED electrical addressing and on carrier injection in the structure.

There are very few state-of-the-art examples of inverted diodes compared to direct diodes. Indeed, the cathode-organic semiconductor interface is very difficult to achieve and control, whereas cathode-organic semiconductor configuration is simpler to realize.

8.5.2. *Using the substrate emitting diode and the top surface emitting diode*

Light emission direction also describes an OLED. Photons can be emitted through the substrate which hence must be transparent. On the contrary, photons can be emitted through the top of the stack. This choice implies a discussion on the types of electrode. Usually, for diodes emitting through the substrate, the anode is composed of a TCO (transparent conductive oxide) such as ITO[8].

OLED top surface emission liberates the constraints on the nature of the substrate. Then an addressing pattern can be placed under the OLED stack to create an active matrix display. In this case, a reflective base electrode is then compulsory to optimize the created cavity. At the same time, less conventional substrates, such as steel may be used for this kind of structure and pave the way for large and low cost lighting devices.

8.5.3. *Heterojunction diode and band engineering*

8.5.3.1. *Electrode choice*

To generate photons, carriers must be first injected in the semiconductor material; an anode and a cathode have to be chosen with a work function which limits the potential barrier to respectively the HOMO and LUMO levels of the adjacent semiconductor. Moreover, at least one electrode must be transparent or half-transparent and both must have a sufficient thermodynamic stability to prevent early decay of the component.

8.5.3.2. *Quenching phenomenon*

Assuming alignment of Fermi levels of electrodes on HOMO and LUMO levels (hence an ohmic junction), organic semiconductors have neither identical mobilities nor conductivities. Consequently, in a single layer OLED (as represented on the left of Figure 8.8), the most likely

---

8. ITO: indium tin oxide.

formation area for hole-electron pairs is close to one electrode. It offers a non-radiative relaxation possibility, or *quenching*, for excitons formed in the electrode *continuum* of levels.

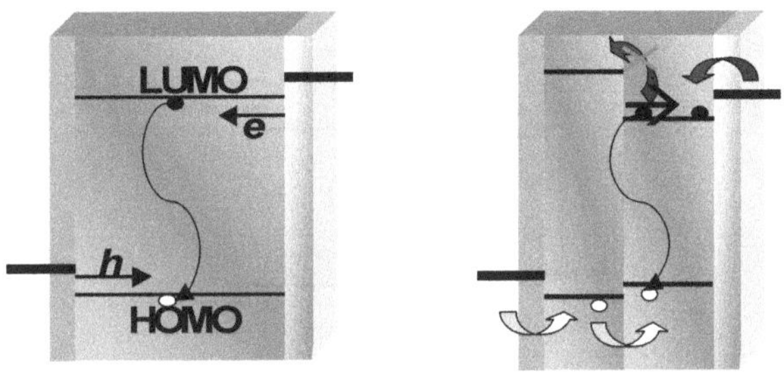

Figure 8.8. *Band diagram for a single layer diode (on the left) and a bilayer diode (on the right)*

Single layer diodes show limited efficiency because of the quenching phenomenon, conductivity difference between HOMO and LUMO levels and also because the injection differences at the electrodes imply domination of a single type of carrier. Indeed, unbalanced charges in the structure necessarily imply a poor final efficiency.

To break free from the quenching phenomenon, the structure of OLEDs can evolve towards a bilayer diode which is composed of two organic semiconductors. Associating a high electron conductivity semiconductor with the cathode and a high hole conductivity semiconductor with the anode allows the recombination zone to be fixed at the heterojunction, far from the electrodes.

8.5.3.3. *Materials specificity*

Today, OLED architectures have evolved considerably toward multilayered structures, using numerous materials according to their major specificity. They are named as follows:

– HTL (*hole transporting layer*): high holes conductivity layer;

– EBL (*electron blocking layer*): low energetic LUMO level layer;

– EL (*emissive layer*): specific color emissive layer;

– HBL (*hole blocking layer*): high energetic HOMO level layer.

– ETL (*electron transporting layer*): high electron conductivity layer.

The multilayered OLED concept is presented in Figure 8.9 where the previously described layers are represented from left to right.

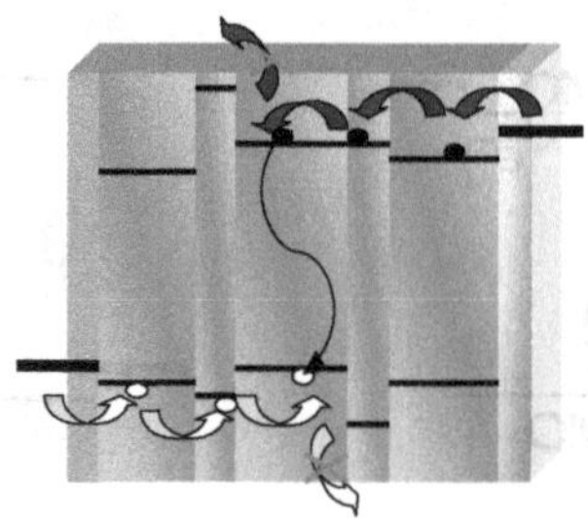

Figure 8.9. *Band diagram of a multilayered OLED with, from left to right: anode/HTL/EBL/EL/HBL/ETL/cathode*

The hole injection layer (HIL) and the electron injection layer (EIL) can also be added to the above list. They improve carrier injection from the electrodes to the transport materials.

8.5.3.4. *Electrical doping*

As in classic microelectronics, an OLED transport layer can be doped but this limits the use of HIL and EIL materials. The doping term is not really appropriate for organic microelectronic; indeed, specific properties molecules are introduced. Thus, for an n-type (p-type) doping, a hole (electron) carrier material has to be used. Its HOMO (LUMO) level has to be between the LUMO (HOMO) level of the EIL (HIL) and the work function of the cathode (anode).

The use of doped transport layers loosens the constraints on the work function of electrodes, makes it possible to increase the thickness of transport layers without raising the working voltage, and raises the density of injected carriers inside blocking and emissive layers.

8.5.4. *Light extraction*

As for inorganic diodes, organic materials usually have an optical refractive index of around 1.8. Emitted photons from the emissive layers are mainly trapped in the guided modes of the stack. Different methods exist that optimize light extraction out of the devices.

Understanding of interferences phenomenon occurring in red, green and blue emitting OLED structures is the key to obtaining high efficiency. This is particularly true for top emissive structures.

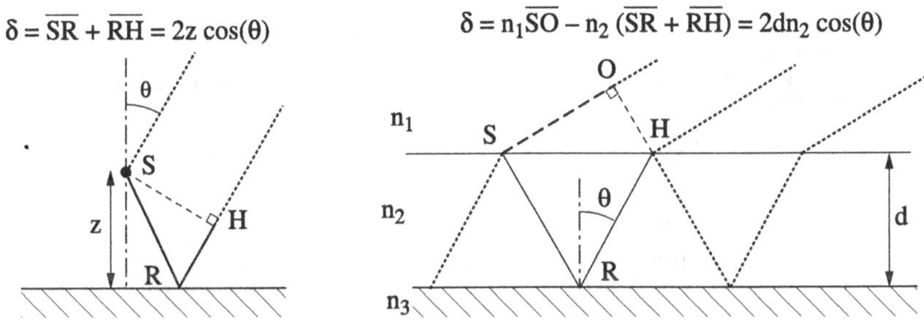

Figure 8.10. *Schematic view of two types of encountered interferences in OLEDs (left: direct interferences, right: cavity effect)*

Two major type of interference mechanisms are presented in Figure 8.10. Their optimization will raise device output coupling, that is, the ability of photons to escape from the OLED. On the right, direct interferences between the most reflecting electrodes and the emitter and the second electrode clearly illustrate the effects of the cavities generated in these structures.

8.5.5. *Fluorescence versus phosphorescence*

Organic luminescent materials can exist in different forms that differ from one another by the way they are implemented or by their radiative relaxation mode.

Indeed some materials called luminescent dopants, in comparison to simple emitters, need to be dispersed in an organic matrix. This allows the transfer of electron-hole pairs towards dopant HOMO and LUMO levels, or transfer of the exciton formed in the matrix towards the dopant (Förster transfer) or even the matrix emitting a photon which is absorbed by the dopant, which in turn re-emits (Dexter transfer).

Moreover, among these materials, we should distinguish between fluorescent material that emits a photon from a singlet electronic level of the material, and phosphorescent material the emission of which results from relaxation from a triplet level (see Figure 8.11).

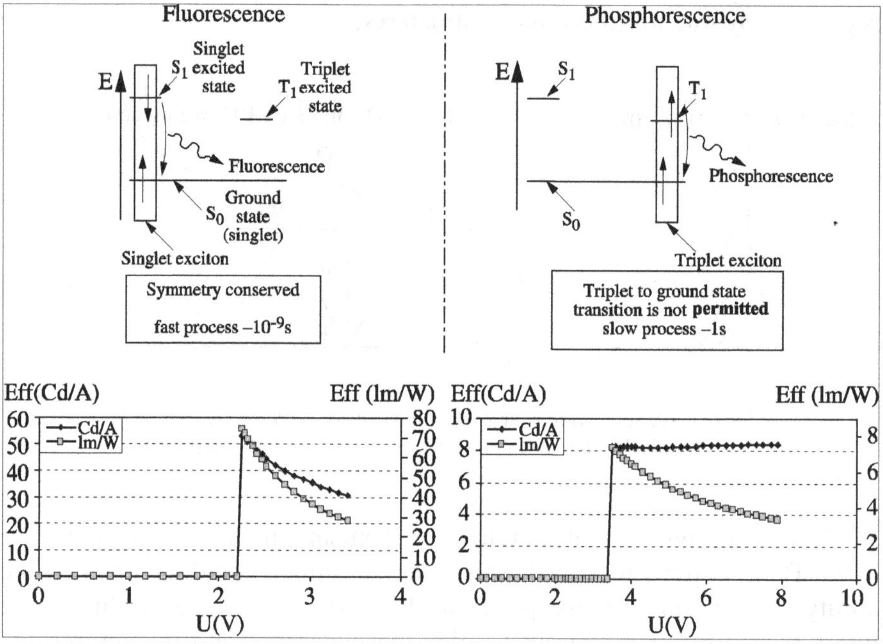

Figure 8.11. *Band diagram (top) and efficiency (bottom) of two luminescent organic material a fluorescent material (left) and a phosphorescent material (right)*

Theoretically, Fermi's golden rule forbids radiative relaxation from triplet levels. Besides, triplet levels are three times more numerous than singlet levels and consequently reduce to 25% the maximum theoretical quantum efficiency of OLEDs. However, for molecules exhibiting a strong *spin-orbital* coupling, such as organo-metallic with a metallic center of atomic weight, the mix between triplet and singlet states allows radiative relaxation from triplet levels and considerably improves electroluminescent efficiency. A very well known example of a phosphorescent small molecule is Irppy[9], a green luminescent dopant.

9. Irppy: tris (2-phenylpyrridine) iridium.

Very often, especially for phosphorescent emitters, the emissive layer is composed of two materials: the matrix or host and the luminescent doping material. This allows decorrelation between specific properties of each material. HOMO and LUMO levels in the host contribute to a reduction of ohmic losses and energy transfer towards the doping material. Two types of transfer can be distinguished: Forster and Dexter.

Of course, phosphorescent-based devices have a longer response time than devices with fluorescent emitters, due to relaxation from a triplet level in the phosphorescent case. Besides, as shown in Figure 8.11, phosphorescent materials lead to a very high conversion efficiency in Cd/A or Lm/W at low voltage. However, these ratios decrease quickly with increasing voltage, due to the triplet-triplet annihilation phenomenon.

Consequently, according to the device requirements such as active or passive matrix display, classical lighting or high luminance display, we should choose the technology that is best suited to the need at hand. For low-resolution displays with passive matrix, fluorescent molecules are a better choice, whereas for active matrix displays and classical lighting, phosphorescent materials should be chosen. Currently, phosphorescent materials are not yet available under polymeric form and for blue emission of small molecules.

8.6. OLED lighting dedicated architectures

The goal of lighting is to imitate the solar spectrum, that is, to generate a luminous spectrum corresponding to the sun in the visible range, from 400 nm to 800 nm. To do so, several emitters must be associated to obtain a large spectral coverage and a "pure" white color. White emission may be defined by different parameters such as chromatic coordinates, the color rendering index and color temperature.

8.6.1. *Single emitting layer structure*

Single emitting layer structure requires a concentration of two or more emitting materials in the same organic material forming the single layer. However, co-evaporation with more than two materials is a difficult process to control and to reproduce, especially at low doping levels. That is why it is rarely used.

In the case of two emitting materials in the same layer, the following condition must be fulfilled. The chromatic coordinates of their emitting spectrum must be linked by a straight line including the desired white point coordinates, e.g. x=0.33 and y=0.33 for the D33 white point. Moreover, the chromatic coordinates of the resulting emitter, that is, the position on the linking segment between the two spectrums, depends on the ratio between the two materials.

However, perfect control of each dopant concentration is mandatory because there is necessarily a competition between emitters with different gaps. Indeed, from a thermodynamics point of view, lower energy gaps are preferably de-excited. Figure 8.12 presents the evolution of chromatic coordinates for a small molecule-based OLED with a PAP-NAP host (blue emitter) and according to the concentration of a rubrene dopant (yellow emitter).

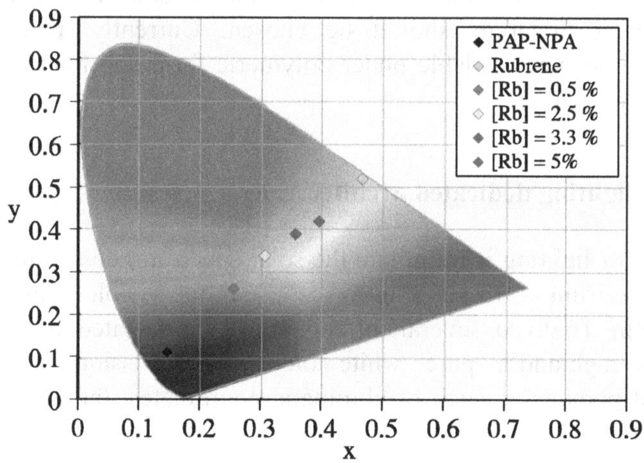

Figure 8.12. *Evolution of chromatic coordinates at fixed voltage according to the concentration of yellow emitting rubrene dopant for a PAP-NAP-based OLED (see color plate section)*

Each emitting material concentration is in fact easily controlled thanks to polymer properties. Indeed, components are mixed together before the thin layer deposition process.

The introduction of several emitting centers on the same molecule allows more efficient control of the different relaxation methods according to the gap; indeed, a molecule can only withstand one hole-electron pair.

On the other hand, the mere coexistence of several emitters leads firstly to the filling of lowest energetic gap material levels and hence the evolution of the chromatic coordinates of the diode according to the number of injected carriers.

The highest efficiencies are usually obtained with this simple structure, yet from a quality point of view, the CRI[10] is low.

The introduction of a higher number of emitters (greater than 2) follows the same argument, but is more difficult to put into practice for small molecules, since it requires co-evaporation with control of concentrations.

8.6.2. *Double emitting layer structures*

Double emitting layers ease diode fabrication (notably for small molecules technology) and open the way to a larger number of emitter materials.

Moreover, the hole transport layer is usually partially doped on a thickness close to the emitting layer. This trick considerably helps to limit ohmic losses due to thickness increase in such multi-emitting layers structures.

The main advantage of a more complicated emitting zone lies in the possibility of controlling recombination ratio in the different layers either by choosing appropriate HOMO and LUMO levels of the host containing emitters or by inserting carrier blocking layers.

Thickness increase and, often, induced carrier barriers usually generate a lower efficiency and an increase in driving voltage. On the other hand, a greater number of emitters lead to a better CRI.

---

10. Color rendering index.

8.6.3. *n-emitting layer structures (n≥3)*

Structures with n-emitting layers present the same advantages and drawbacks mentioned before for double emitting layers. The main drawbacks are an increase in driving voltage and a decrease in global efficiency; however, these structures present a better control of carriers' recombination zones and hence a better control of chromatic coordinates, particularly when using specific carriers blocking layers between emitters.

8.6.4. *Stacked OLEDs and tandem structures*

One technique to obtain a white emission consists, for a top surface emissive device, of stacking on the substrate a red, a green and then a blue OLED (in the reverse order for a through substrate emission). This is the preferred order to prevent shorter wave photons from being absorbed by the other emitting materials before getting out of the component. These structures offer effective and active control of the emission of each light component since they are adjustable using the current applied to each diode [SUN 05].

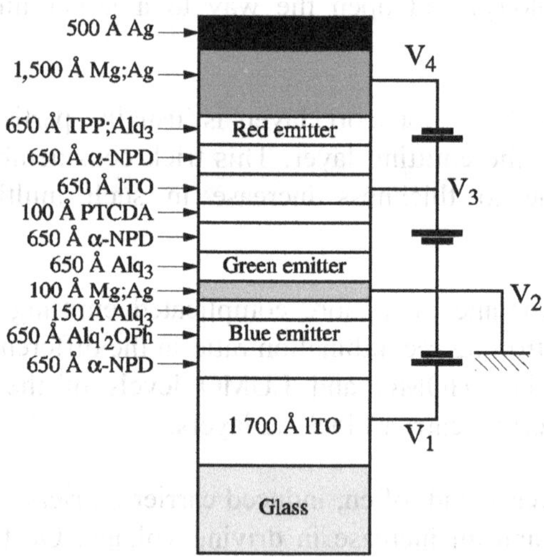

Figure 8.13. *Example of an OLED stack (Stacked-OLED or SOLED [SHE 97])*

There is a very similar structure called tandem OLED (Figure 8.14). It is composed of a stack of two or more diodes with the identical structure and separated by a charge generation material [TER 05]. This stack only has two electrodes, no matter how many diodes are piled up, and hence the OLED current efficiency can be increased [LIA 04]. The major difficulty when using these structures is to resort to transparent electrodes being at the same time the top OLED anode and the bottom OLED cathode (vice versa if the structure is inverted).

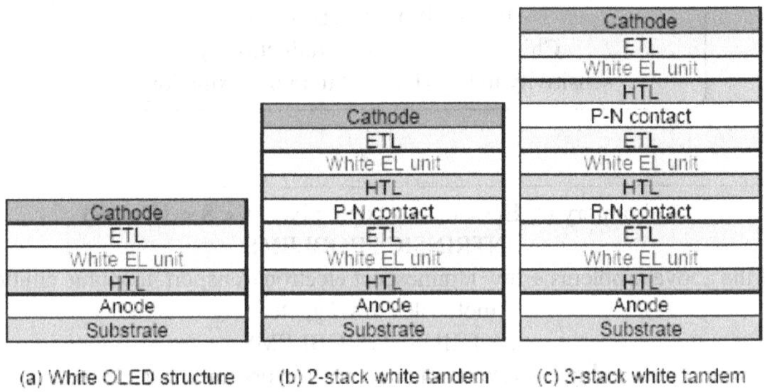

Figure 8.14. *Comparison between a simple white OLED structure (a) and two structures of the tandem type (b) and (c) [HAT 06]*

8.6.5. *Converters (down conversion)*

To limit the number of emissive materials, some structures use a color converter, which is a layer of material converting high energy photons into lower energy photons [KRU 06].

8.7. OLED stability and lifetime: encapsulation issue

The issue with organic materials is their high sensitivity to gaseous elements in the atmosphere, particularly O_2 and H_2O. This is why their protection in an electroluminescent device is important to prevent accelerated decay. Generally, a protection, or an encapsulation, of circuits is performed in order to protect active organic layers and sensitive cathode metal. Figure 8.15 describes the encountered problems with organic semiconductors and depicts the encapsulation concept for the OLED (this concept is also valid for all electronic devices using hydrocarbonated materials such as photovoltaic cells or transistors). As illustrated in the figure, it is mandatory to protect the organic layer at each fabrication step of the assembly, that is, from the rough material obtained by chemical synthesis (safeguarding of the powders in an inert atmosphere for example) to the commercial device (the case of a lamp or display).

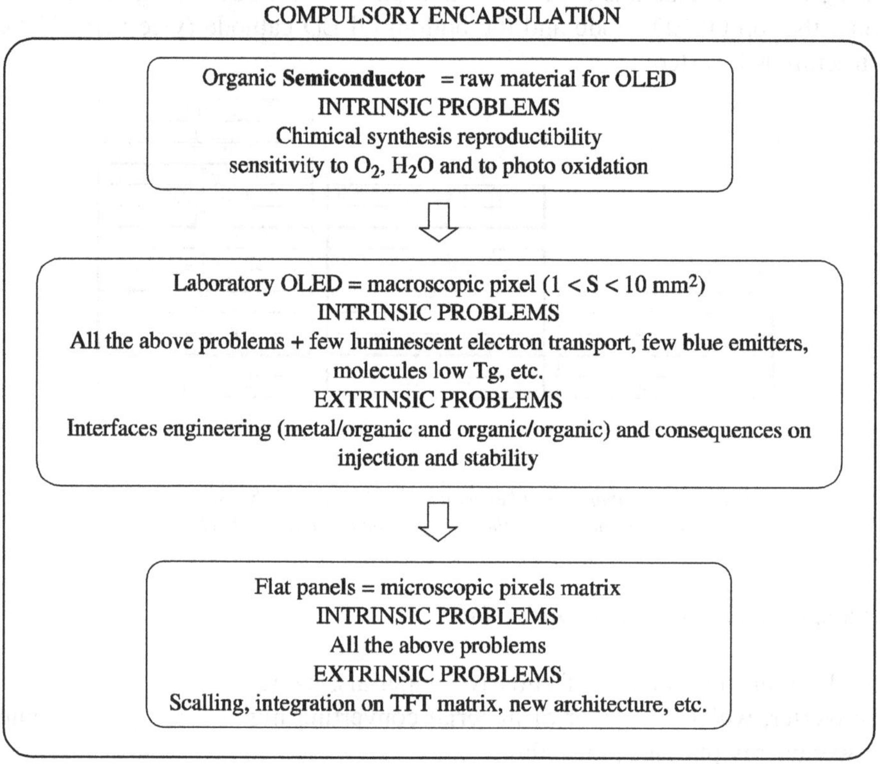

Figure 8.15. *Encapsulation issues for OLED devices*

The first encapsulation concept of an organic diode was published in 1994 [BUR 94]. The pixel was built on a glass/ITO substrate. On the other side it was protected by a glass slice. The assembly was held together thanks to an airtight resin. In 1996 this concept was patented by Pioneer who added a desiccant (P2O5) inside the space housing the device [PIO 96]. Today it is still the preferred process that can be encountered in commercial products.

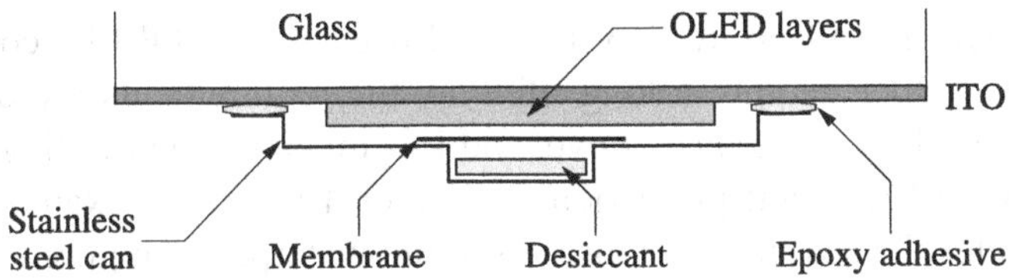

Figure 8.16. *OLED device first encapsulation concept (source: Pioneer 1996)*

In addition, the main technological breakthrough for OLED devices is the ability to develop products (displays, lamps) on flexible substrates; some expert are even asserting that there lies their future commercial viability. Plastic (PET, PEN) or very thin steel substrates are then used (Figure 8.17).

Figure 8.17. *OLED 3" passive display on a PET plastic substrate (source: Pioneer 2002) (see color plate section)*

However, to keep the major interest of these applications, organic devices should be encapsulated without stiff parts. Consequently, monolithic encapsulation solutions were developed. The monolithic term refers to the use of thin layer barriers deposited on the circuits by PVD or CVD methods, to protect them against atmosphere attacks.

This is an old concept, introduced a long time ago in the food packaging industry. As an example, to protect biscuits packaged in plastic packets, these packets are wrapped in aluminium thin films generally deposited by a

PVD process in roll to roll systems. However, for OLED commercial applications, which require a long lifetime, the requirements are drastically different from those described above, and it is much more gruelling to obtain the suitable and resistant protection for the OLED. As an example, Figure 8.18 illustrates the scaling phenomenon of the waterproof required capability according to the desired application. Organic electronic technology is the one which requires the highest waterproof qualities of the encapsulating layers in order to provide devices reliable for several years. They are referred to as ultra high barriers.

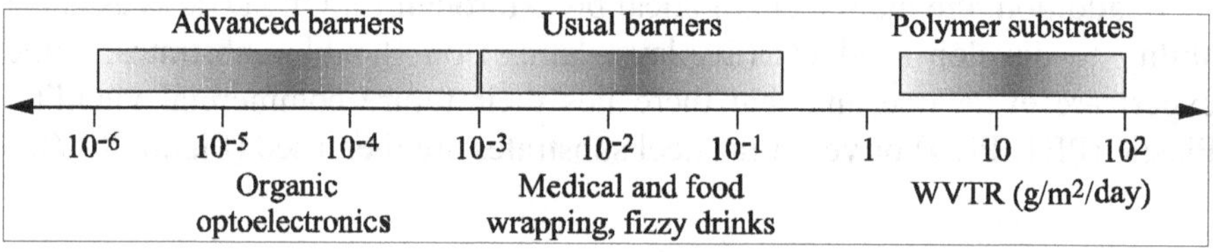

Figure 8.18. *Scaling phenomenon of the waterproof required capability according to the application [CRO 07]*

8.8. OLEDs for lighting

Today, OLEDs are not yet introduced in the lighting market because devices do not reach the required levels of performance to compete with inorganic solutions. Nonetheless, there are still good and promising achievements in some leading sectors of the industrial lighting domain. As an example, in Figure 8.19, there is a panel created by General Electric, lighting up a large surface. This kind of source has been shown by Philips and OSRAM, but there is still a long way to go before compliance with the requirements of this kind of application is achieved.

Figure 8.19. *OLED-based lighting panel (source: General Electric 2005) (see color plate section)*

The following table gives an overview of the foreseen evolution of OLED technology from 2004 to 2013 in order to obtain a commercially viable lighting device [BAR 04]. Besides planned improvements, performances such as efficiency and lifetime have been added to the evolution of the device structure. Hence, great technological advances should allow the fabrication of lighting organic panels on flexible substrates in around 2013. However, the issue of fabrication cost will inevitably be raised. In this case, roll to roll processes on plastic are envisioned for OLED fabrication. This is an asset for OLED technology, since other technologies are not eligible for this kind of process. It is also its Achilles' heel since industrial production of high volumes in order to produce organic thin layers that require fabrication under inert atmosphere seems difficult. Besides, plastic substrate permeability will inevitably be an issue if fabrication on this kind of substrate proves its feasibility. Robust encapsulation technologies will thus have to be developed to reach a protection against atmosphere in the range of 10^{-6}g/m^2/day.

Properties	Step 1	Step 2	Step 3	Step 4
Year	2004	2007	2010	2013
External quantum efficiency (%)	5	12.5	20	30
Luminous efficiency (Lm/W)	20	50	80	120
CRI	75	80	85	90
Lifetime at 2,000 cd/m^2	10 K	20 K	40 K	50 K
Lightening panels width (cm)	36	100	100	> 100
Panels thickness (mm)	2	1	0.5	0.5
Panels weight (g/cm^2)	0.5	0.25	0.1	0.1
Fabrication cost ($/m^2)	120	60	40	30
Device structure	A/OLED/A	A/OLED/B	A/OLED/B	C/OLED/B

Table 8.1. *OLEDs for lightening planned evolution over the 2004-2013 period. Peculiar substrates evolution – A: glass; B: monolithic encapsulation; C: flexible subtrate with monolithic encapsulation*

8.9. Bibliography

[BAR 04] BARDSLEY J.N., IEEE, "International OLED Technology Roadmap", *Journal of Selected Topics in Quantum Electronics*, vol. 10, p. 3-9, 2004.

[BAS 81] BÄSSLER H., "Localized states and electronic transport in single component organic solids with diagonal disorder", *Physica Status Solidi* (b), vol. 107 p. 9-54, 1981.

[BLO 96] BLOM P.W.M., DE JONG, M.J.M., VLEGGAAR, J.J.M., "Electron and hole transport in poly(*p*-phenylene vinylene) devices", *Applied Physics Letters*, vol.68, p. 3308, 1996.

[BOR 98] BORSENBERGER P.M., WEISS D.S., *Organic Photoreceptors for Perography*, Marcel Dekker, New York, 1998.

[BUR 90] BURROUGHES J.H., BRADLEY D.D.C., BROWN A.R., MARKS R.N., MACKAY K., FRIEND R.H., BURN P.L., HOLMES, A.B., "Light-emitting diodes based on conjugated polymers", *Nature*, vol. 347, p. 539-541, 1990.

[BUR 94] BURROWS P.E., FORREST, S.R., "Electroluminescence from trap-limited current transport in vacuum deposited organic light emitting devices" *Applied Physics Letters,* vol. 64, p. 2285, 1994.

[CHE 99] CHEN B.J., LAI W.Y., GAO Z.Q., LEE C.S., LEE S.T., GAMBLING W.A., "Electron drift mobility and electroluminescent efficiency of tris(8-hydroxyquinolinolato) aluminum", *Applied Physics Letters*, vol. 75, p. 4010-4012, 1999.

[CON 56] CONWELL E.M., "Impurity Band Conduction in Germanium and Silicon", *Physical Review,* vol. 103, p. 51, 1956.

[CRO 07] CROS S., "Propriétés barrières des polymères utilisés en emballage", *Techniques de l'ingénieur*, vol. AM 3, p. 160-1, 2007.

[CRO 98] CRONE B.K., CAMPBELL I.H., DAVIDS P.S., SMITH D.L., "Charge injection and transport in single-layer organic light-emitting diodes", *Applied Physics Letters*, vol. 73, p. 3162, 1998.

[GIL 72] GILL W. D., "Drift mobilities in amorphous charge-transfer complexes of trinitrofluorenone and poly-*n*-vinylcarbazole", *Journal of Applied Physics,* vol. 43, p. 5033, 1972.

[HAT 06] HATVAR T.K., SPINDLER J.P., VAN SLYKE S.A., "Low-Voltage White Tandem Structures for Fabricating RGBW AMOLED Displays", *SID 06 Digest 70.2*, vol. 37, p. 1964-1967, 2006.

[IOA 98] IOANNIDIS A., FORSYTHE E.W., GAO Y., WU M.W., CONWELL E.M., "Current–voltage characteristic of organic light emitting diodes", *Applied Physics Letters*, vol. 72, p. 3038, 1998.

[KRU 06] KRUMMACHER B.C., CHOONG V.E., MATHAI M.K., CHOULIS S.A., SO F., JERMANN F., FIEDLER T., ZACHAU M., "Highly efficient white organic light-emitting diode", *Applied Physics Letters*, vol. 88, p. 113506, 2006.

[LIA 04] LIAO L.S., KLUBEK K.P., TANG C.W., "High-efficiency tandem organic light-emitting diodes", *Appl. Phys. Lett.*, n°84, p.167-169, 2004.

[MAR 62] MARK P., HELRICH W., "Space-Charge-Limited Currents in Organic Crystals", *J. Appl. Phys.*, *Journal of Applied Physics,* vol. 33, p. 205, 1962.

[MIL 60] MILLER A., ABRAHAMS E., "Impurity Conduction at Low Concentrations", *Physical Review*, vol. 120, p. 745, 1960.

[MOT 56] MOTT N.F., "On the transition to metallic conduction in semiconductors", *Canadian Journal of Physics*, vol. 34, p. 1356, 1956.

[PIO 96] PIONEER, Patent EP0776147, 1996.

[POP 63] POPE M., KALLMAN H.P., MAGNATE P.J., "Electroluminescence in Organic Crystals", *Journal of Chemical Physics*, vol. 38, p. 2042, 1963.

[ROB 79] ROBERTS G.G., MCGINNITY M., BARLOW W.A., VINCETT P.S., "Electroluminescence, photoluminescence and electroabsorption of a lightly substituted anthracene langmuir film", *Solid State Communications,* vol. 32, p. 683, 1979.

[SAV 97] SAVVATE'EV V.N., TARABIA M., CHAYET H., FARRAGI E.Z., COHEN G.B., KIRSTEIN S., DAVIDOV D., AVNY Y., NEUMANN, "Space-charge limited current and electroluminescence in conjugated polymer-based LEDs", *Synthetic Metals*, vol. 85, p. 1269-1270, 1997.

[SCO 00] SCOTT J.C., BROCK P.J., SALEM J.R., RAMOS S., MALLIARAS G.G., CARTER S.A., BOZANO L., "Charge transport processes in organic light-emitting devices", *Synthetic Metals*, vol. 111-112, p. 289-293, 2000.

[SHE 97] SHEN Z., BURROWS P.E., BULOVIC V., FORREST S.R., THOMPSON M.E., "Three-Color, Tunable, Organic Light-Emitting Devices", *Science,* vol. 276, p. 2009-2011, 1997.

[SUN 05] SUN J.X., ZHU X.L., PENG H.J., WONG M., KWOK H.S. "Highly Efficient Stacked OLED Employing New Anode-Cathode Layer" *SID, Digest P-129*, vol. 36, p. 797, 2005.

[TAN 87] TANG C. W., VAN SLYKE S.A., "Organic electroluminescent diodes", *Applied Physics Letters*, vol. 51, p. 913, 1987.

[TER 05] TERAI M., KUMAKI D., YASUDA T., FUJITA K., TSUTSUI T., *Current Applied Physics*, vol. 5, p. 341-344, 2005.

[VIN 82] VINCETT, P.S., BARLOW W.A., HANN R.A., ROBERTS, G.G., *Thin Solid Films*, vol. 94, p. 171, 1982.

List of Authors

Jean-Michel DESWERT
Laborelec
Belgium

Amélie DUSSAIGNE
Ecole Polytechnique Fédérale de Lausanne (EPFL)
Switzerland

Christian EUGÈNE
Katholieke Universiteit Leuven
Belgium

Marc FONTOYNONT
National School of State Public Works
University of Lyon
France

Adrien GASSE
CEA-LETI
Grenoble
France

Philippe GILET
CEA-LETI
Grenoble
France

Nicolas GRANDJEAN
Ecole Polytechnique Fédérale de Lausanne (EPFL)
Switzerland

Tony MAINDRON
CEA-LETI
Grenoble
France

Philippe DE MIERRY
CNRS-CHREA
Sophia-Antipolis
France

Patrick MOTTIER
CEA-LETI
Grenoble
France

David VAUFREY
CEA-LETI
Grenoble
France

Françoise VIÉNOT
Muséum National d'Histoire Naturelle
Paris
France

Georges ZISSIS
LAPLACE Laboratory
University of Toulouse 3
France

Index

Printed in the USA/Agawam, MA
September 16, 2026

910661.025